Fungus Theory of Conscious-Growth

An Analysis of Life's Formation Throughout the Universe

Mark L Christensen

ISBN: 978-1-7334555-1-0

DEDICATION

Dedicated to my father and mother
Larry and Sherry Christensen

CONTENTS

AUTHOR'S NOTE

If we are here for a limited time and generate so much emotional energy, thought, concern, hope, faith, endurance, stride, scream, kick, kill, and destruction only to later pass away, then what's it all for? Why are we here?

The process of creating this book has been fungal to say the least. It has formed itself from an eight-year intense research process using the least amount of discretion possible. Taking myself out of the equation has been a personal endeavor and has reformed my enforced and new self-identity.

My own evaluation of all the material presented is as hard as it is to admit: all actions carried out by not only humans but every animal including the single-celled, and that means every action is an emotional step forward in the process of growing a consciousness to its maximum growth state, whatever that is. With that process in mind the Earth is "The perfect machine."

FOREWORD

Fungus Theory of Conscious Growth – The Idea Starts

In 1970, as a nine-year-old boy, I was watching fireworks on the Fourth of July. As I watched them rise and explode in the night sky, I marveled at how each one looked like a Milky Way galaxy for a few seconds. I thought that maybe the fireworks explosions were like a miniature Big Bang like our universe, and the sparks spreading outward were the swirls of miniature stars. In those sparks, I imagined quickly-formed stars that then died out, in the same way the universe was created. The fireworks had formed a temporary little universe right in front of my eyes, and in an instant, all the sparks vanished into total darkness.

I then imagined that each one of those sparks could have quickly developed tiny little planets that rotated around it, perhaps even giving life to an entire human species, and then vanishing in the blink of an eye.

I then slowed down that timeframe in my head and compared it to human beings on earth rotating around a spark from an exploding fireworks show we call our Sun, a very big spark indeed; the sun being a more gigantic version of the fireworks spark I was observing. And just like the brief life of a firework, our lives on Earth may be a mere instant in cosmic time. Then I took it a step further by thinking that in every firework that ever exploded, there were worlds developing inside of them watching other fireworks, and inside those were more worlds exploding, and so on and so forth. Kind of like an infinity mirror.

That memorable day had me thinking about trying to answer these questions: if Earth and other planets in our universe are short-lived, then what is it all for? What is the world we are actually in? Is life an accident? Why is life temporary? Why is my participation seemingly so significant, if it is all just for a short time on a planet short-lived?

In summation of this story, why is it that in the course of our lives, we are forced to go through so much pain and loss? It has to be that human life is much more significant than a simple accident. In my earlier years looking at the world I was placed in, not having the answers to my temporary existence frustrated me and I wanted to get to the bottom of it. All of life's endeavors – struggles, development, growth, education, growing families, saving lives – just to inevitably die didn't make sense. I couldn't accept that it's all for nothing in the end. Is that what the universe wants? The pain people go through in life is real. Physical and mental struggles, accomplishments, strides, sacrifices, dedication, blood, sweat, and tears have to have a purpose! It made me conclude that my existence must be for something bigger than me. What is that something? It angered me to the point that from that day on, everything I would encounter as an experience in my life would have this question behind it. What is this all for?

In the simplest form it said: starting with bacteria feeding on minerals as a single cell evolved into a fungus called slime mold. This slime mold is both equally plant and animal about 2 BYA. After 500,000 years, this fungus evolved to become multicellular; then it split. The plant half of the fungus split away and went onto land to form mushrooms and the root system of plant life we see today. The animal/fungus side went on for a few hundred million years after which it split off again, leaving the fungus to evolve as the kingdom we see today. The animal fungus evolved in the ocean to develop a vertebrae and a head. Both plant and animal fungus worked separately yet simultaneously adapting to Earth's composition developing all life forms needed to eventually stand up. When the animal fungus eventually stands up on land, it walks over, picks a tomato, and eats it. This demonstrates that the same original fungus is eating itself. Both mankind and vegetables had evolved last, making this a planned system. This imposes two questions: 1. Why this method of growth? 2. What for?

Why this method of growth? To Hide True Origin (HTO) from what may become conscious enough to detect its True Origin. This means that if that consciousness can find or detect its True Origin, it would no longer participate in the growth of its own consciousness. A catch 22, looking more like a design for conscious growth by a conceiver with the intent to grow it. It's as though the conceiver understands that what may evolve to become smart enough to find its True Origin, that growth would then not meet its highest expectations of growth, making this a system of HTO.

What for? To slowly build a brain capable of a conscious state of awareness, and the growth of that consciousness to its maximum, or Maximum Conscious Growth (MCG). Conclusion: Because mankind has the technological means to leave Earth forever, it supports the evidence that man is the spore of this very large single fungal growth.

INTRODUCTION

This book is a collection of facts, evidence, and events that support the thesis that all life generated on Earth is from a very large and very old "*fungus*" of which humanity is ultimately the spore. The following facts and evidence support this theory.

1. The evolutionary split in the original fungus on Earth around one billion years ago that formed both plant and animal is one large system today. The original fungus continued on and is the fungus kingdom we have in present time.

2. All plants and animals retain certain fungal characteristics from that split.

3. The behavior of all plant and animal life throughout evolution is highly comparable to the growth of fungi in slime mold and the mushroom life cycle.

For instance, at the end of its cycle, this particular fungi system, a mushroom, produces spores to procreate and exponentially expand its new life. Since the Earth will ultimately end, humanity's purpose in this system is comparable to that of the spore of Earth's ancient "*fungus*."

Ultimately, this "spore" will be forced from the Earth's overheating to leave and continue its growth in the cosmos. Thus, all life on Earth is one ever-evolving "*fungus*" that has developed humanity as its spore to continue life beyond the bounds of Earth.

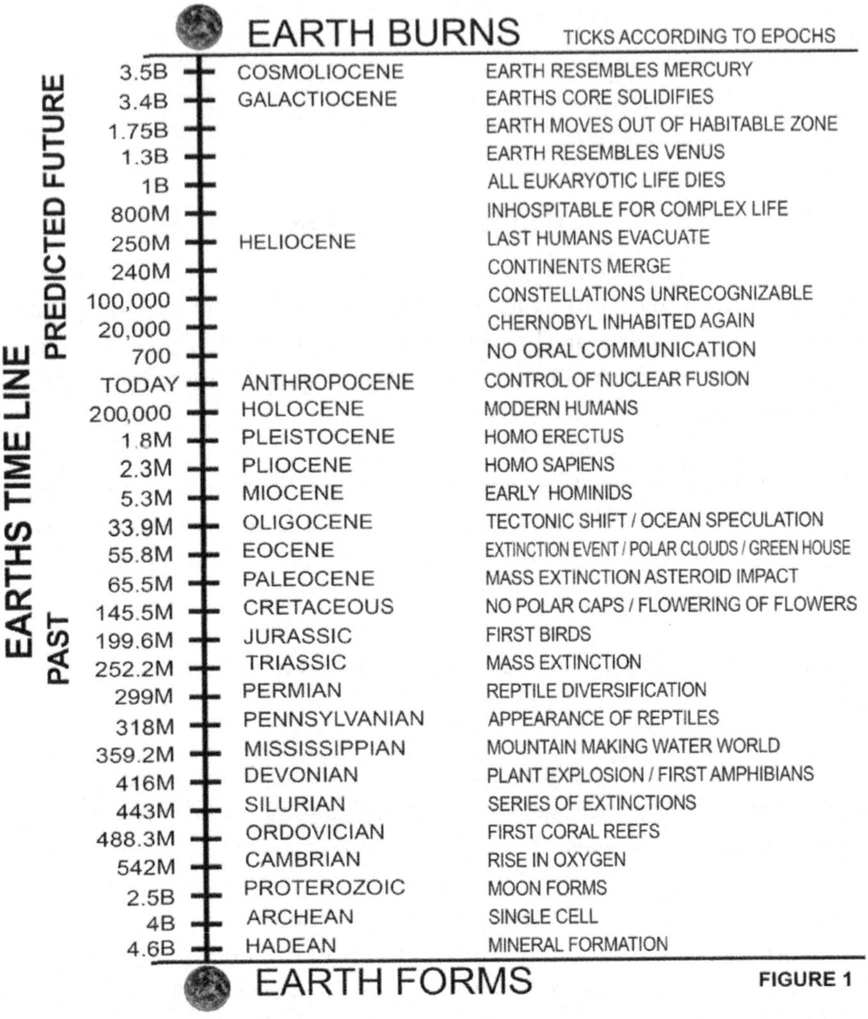

Figure 1: Earth's timeline

This figure represents the biological evolution of a *fungus* from beginning to end. It shows the many different phases the *fungus* will go through in order to spawn, generate and launch a conscious-growth. From bombardment to cooling and then reheating the biosphere in which that *fungus* is born and destroyed, this continually evolving environment will produce a matured consciousness that will be propelled into outer space to regenerate.

This is a system of absolutes. Under such conditions, it would seem the same process of conscious-growth of organic life should be replicated in the same manner elsewhere throughout the universe.

Viewing this process, from humble terrestrial beginnings to extraterrestrial existence in its entirety, as a system that brings a nascent conscious-growth to its maximum potential reveals the function, and arguably the purpose, of the universe: that being as a mechanism for conscious-growth.

The mere fact that you can read this manuscript today, while humans were incapable of storing and transferring information–via formal methods of communication like writing–thousands of years ago, is evidence that human life, and all life on Earth that was or will be, is a continuing conscious-growth process.

This demonstrates that a higher level of consciousness exists today in comparison to that of past times. And that as time moves forward, consciousness increases yet at a faster rate still. More recently, even over short periods of time, human consciousness has grown significantly. This demonstrates that such conscious-growth occurs at exponentially accelerating rates. We hold more power in our hands now within a common smartphone than what was used to get the astronauts to the moon in 1969, and this giant leap for mankind has taken place within less than the span of 100 years.

With only this evidence presented to you, it would seem like the craziest luck or accident of all accidents that this would occur in the universe by fluke or chance. It would make more sense if this was an intended result or a logical consequence of some underlying mechanism.

This work aims to demonstrate that conscious-growth occurs due to underlying principles inherent within the universe, principles as basic as gravity. These limitations force continual adaptation, which is best displayed in the evolution of life forms and their response to their environment. An example of such a forced or enforced adaptation that is quintessential for any conscious-growth to overcome is the death of its biosphere and as stated before, Earth's fate is no different. We are becoming more and more aware of Earth's inevitable demise in modern times due to scientific advancement in the understanding of the sun's increasing heat output. Knowledge of this fact leads to the realization that Earth will overheat and become inhospitable to life at some point in the future.

The Human Condition of Absolutes

This "Human Condition of Absolutes" causes fear and concern over the longevity of our species as a whole. The awareness of this condition is only beginning to permeate the general public's awareness and these ramifications are personal to us. However, this is not the case since this mechanism is at play in every star system throughout the universe. Thus, this is a universal limitation for any life to form, become sentient, and evacuate its birthplace. Ultimately, the proof that this mechanism exists will be shown to be evident in the formation and evolution of life itself. Thus, it will become clear to the reader that the product of the universe is Maximum Conscious-Growth (MCG) and it is achieved by the mechanisms enforced by the very physics that governs the universe itself. The physical manifestation of this conscious-growth is the ultimate appearance of a planetary "*fungus*" that produces a human spore for eternal space life.

If it is an absolute that this planet's ability to continue reproduction for all the life that may evolve on it will eventually end, then, this is a system by which a product must result. That product would be a spore-like life-form that may live beyond its post-terrestrial origins and then continue to proliferate in the cosmos. We'll call this absolute "The Human Condition of Absolutes" in order to get a spore to succeed its biospheric birthplace.

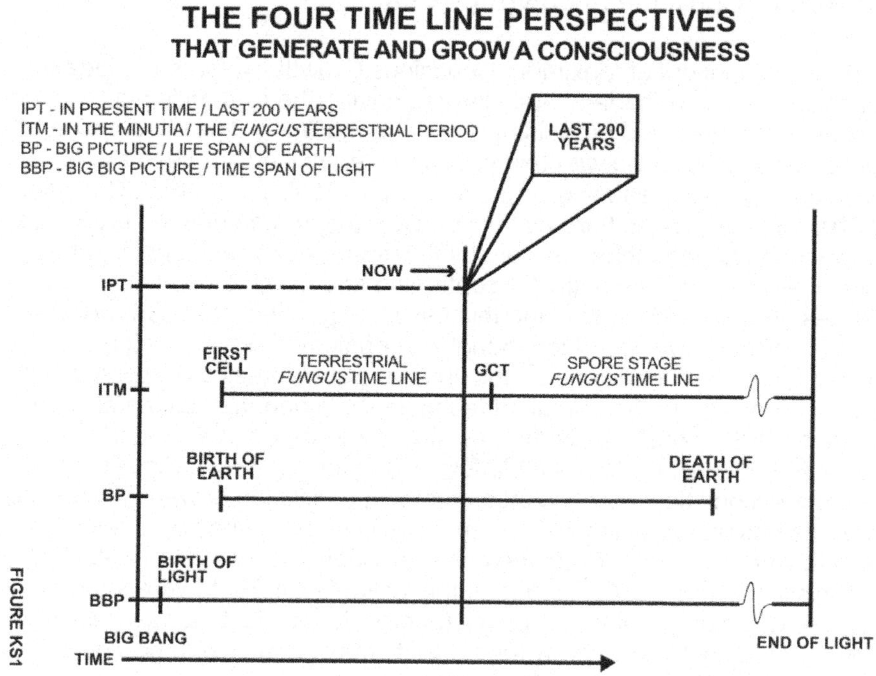

Figure KS1 (The Four Timeline Perspectives)

This figure shows the four different timelines that work simultaneously together that generate and grow a consciousness. This is done in a cold dark vacuum.

The evolution of matter and the numerous changes it has made since it started as heat encompasses the theme and theory that what may develop as intelligence might detect its True Origin. By observing the behavior of all matter, as we see it today, lends the theory that it could be the design itself that makes and suggests that there is some intent to hide True Origin from consciousness as it grows and in a sense the intent to make it. It may be that without this naivety, any biospheric conscious-growth might not be willing to continue forward. This shows in the randomness of matter, not only random but random variable displacement of matter and only structural evidence being that matter had a center point from which it expands and knowing nothing before that conception.

Maximum Conscious-Growth (MCG)

TL;DR The concept of Maximum Conscious Growth explores the idea that consciousness is a fundamental element driving the existence and evolution of matter and organisms since the Big Bang. The observation of consciousness evolving with expanding matter indicates a need for continuous growth, reaching its maximum potential. The Human Condition of Absolutes serves as hard evidence, highlighting the enforced growth of consciousness and technological advancement. Acknowledging layered cognitive realities is essential for confirming and building upon new realities. The conditions shaping this psychology reveal that consciousness is an intended growth product, though the origin remains a mystery. Accepting human conditions of absoluteness contributes to the conscious growth process and the pursuit of technological conquest, ultimately aiming to leave Earth or burn up with it. The key elements include expanding matter in dark cold space as a platform for organics, the inevitability of death, and conditions of absolutes enforcing conscious growth. That being said, the observation also shows the veracity in that behavior to need to grow to its maximum. If Earth were to end today, then we could calculate that this biosphere MCG stopped at this point. But if Earth were to be a successful biosphere and its conscious-growth survive its terrestrial end, then it would continue to grow further to its maximum growth potential, whatever and wherever that would be.

The Human Condition of Absolutes is the hard evidence that what is at work here is the enforced growth of conscious-growth and its technological advancement. This growth process is calculated through observing the acknowledgment of layered cognitive reality, whether in the individual or collective. For instance, one new reality cannot be made without the firm acknowledgment of the previous ones, and this appears to be the only way to confirm the next or even have the potential to confirm the next reality: a controlled and accurate form of building a psychology, which is a growth, through conditions of absolutes.

The conditions that this psychology is subjected to are the evidence that consciousness is a growth product, and an intended growth product. By whom? It's still a total mystery. Of course, if this growth product was not a total mystery, we would not be here to experience it as a mystery, and we most certainly would grow. Accepting the human conditions of absoluteness your psychology encounters is a major contribution to this conscious-growth process and the necessary technological conquest needed to achieve the goal of leaving the Earth.

The God formula below best describes the observed behavior of all expanding matter and organic evolution. It looks highly unlikely that the production of conscious growth can take place in any other environment than a vacuum after with the ignition of expanding matter which is also the physics that hid true origin, which is the setup for the ignition of organic evolution and the *fungus.* All together, these elements develop a psychotic void in an animal that would eventually evolve to observe it forcing it to grow consciously through time and the continuation of that expanding matter.

The God Formula: V+IOEM+HTO+IOFG+VOP+CG = MCG

V: Vacuum
IOEM: Ignition Of Expanding Matter
HTO: Hidden True Origin
IOFG: Ignition Of Fungal Growth
VOP: Void Of Psychosis
CG: Conscious Growth
MCG: Maximum Conscious Growth

EOS

- Expanding matter in dark cold space as a platform for organics
- Inevitable death
- Conditions of absolute that enforce conscious-growth

EVIDENCE OF SYSTEM IN FUNGUS THEORY

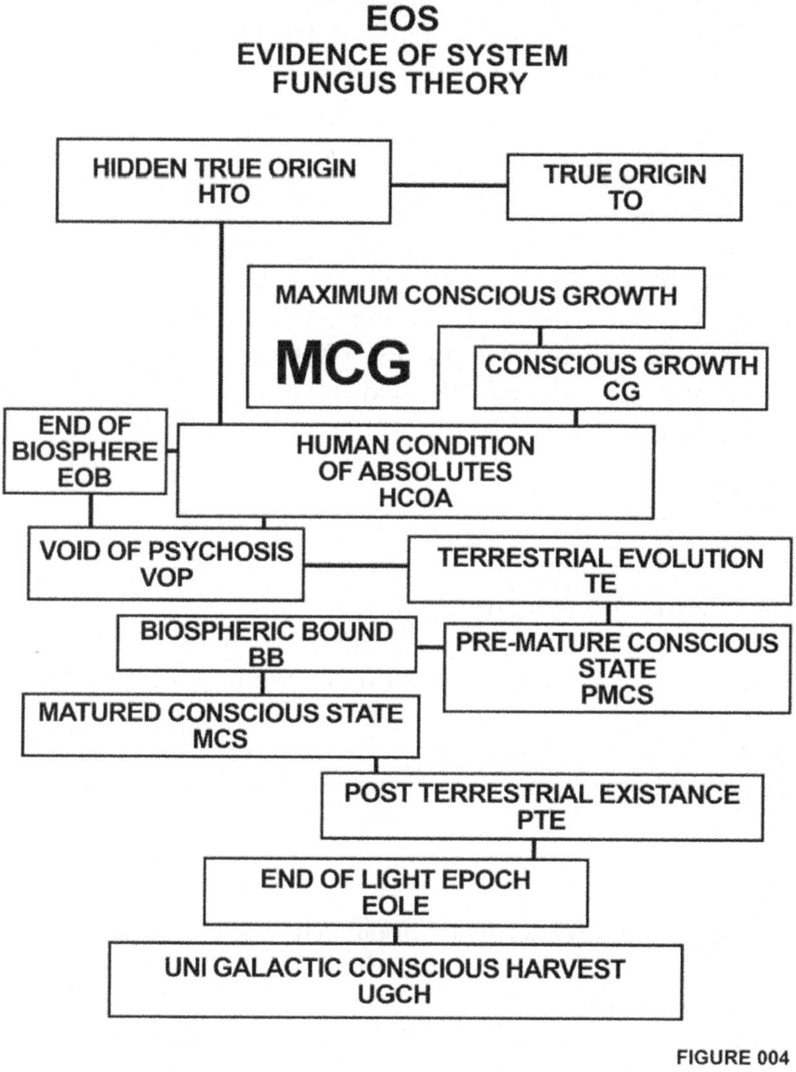

FIGURE 004 (Evidence of System)

This figure demonstrates the use and connection of the elements (field-specific terms and acronyms) that culminate a conscious-growth process to MCG.

TERMS OF EVIDENCE AND ABBREVIATIONS

THE *FUNGUS*
(Italicized to represent the theoretical *fungus*)
Everything biological or alive on earth, on any biosphere – from the first divided cell to the last surviving animal – is one very large and very old *fungus*.

TL;DR Too Long Didn't Read
Summary of the header.

BBP Big Big Picture
The largest visual perspective by the evidence of the universe we can see and calculate to date. From the Pre-Big Bang to the CMB (Cosmic Microwave Background) and predicted and calculated futures, or all that we can see or not see.

BP Big Picture
LifeSpan of a Biosphere.

HCOA Human Condition of Absolutes
This term represents the ultimate condition that humans will have to overcome: the overheating of the earth due to the sun getting larger.

HTO Hidden True Origin
The fact that humans cannot establish the origins of where they came from.

TO True Origin
The true place from where all life originated.

CG Conscious-Growth
Consciousness being a growth or always growing. The growth and oscillation of new information generated from exploration and discovery that propels the collective consciousness.

EOB End Of Biosphere
When biology can no longer be sustained by its environment.

EOS Evidence of System
When it is relevant that there is enough evidence of math, or physics that show a process or product.

VOP Void Of Psychosis
A void in one's concrete identity where true origin cannot be identified with, causing psychosis. This occurs when the *fungus* has developed a brain capable (bipedalism) of self-awareness and acknowledges the fact that it cannot determine its true original origin.

TE Terrestrial Evolution
Evidence of plant or animal life that evolve on a planet.

ITM In The Minutia
The time span of organic life on a biosphere.

BB Biospheric Bound
Organic life being bound to the limits of its biosphere.

PMCS PreMature Conscious State
A state of consciousness in which full maturity cannot be attained. Referring to terrestrial conscious state as premature and post terrestrial as matured conscious state.

MCS Mature Conscious State
A state of consciousness that has matured, having reached post-terrestrial life.

ACS Adolescent Conscious State
The growth of a consciousness that is still bound to its biosphere.

PCC Permeated Collective Conscious
When the majority of a collective conscious growth is informed with a new level of understanding.

PTE Post-Terrestrial Existence
When an animal is existing off a terrestrial surface, preferably in an artificial biosphere.

EOLE End Of Light Epoch
A point in time where there is no longer any illuminating light emitted

in the universe.

UGCH Uni-Galactic Conscious Harvest
A point in time when all the conscious-growths are collected throughout the universe.

TEU Total Expanded Universe
A point in time when all matter expanding in the universe has expanded away from all other matter.

BIG VACUUM
The theory that the matter in the universe quickly expanded due to the walls or membrane surrounding it moving outward, not from a bang.

QEM Quickly Expanding Membrane
A membrane surrounding the center mass called the pre-Big Bang element, quickly expanding away, causing a vacuum and the expansion of matter.

ORIGINAL ORIGIN
A place of origin where conscious organic life originated, before that life started organically on Earth.

LIFE DUST
A potential ingredient within the concoction of the primordial fluctuation, a pre or post big bang element, to ensure the aftermath of biological ignition.

TES Technical Evacuation Standards
The necessary technologically evolved requirements when mankind can leave earth and sustain itself for eternity.

ESS Eternal Self Sufficiency
When mankind can live by its means post terrestrial by relying on its own resources non terrestrially.

TEMPERATURE FUNNEL
The final phase of the organic timeframe on Earth when the temperature slowly increases, forcing humanity to self-genetically mutate and develop technologies to survive and leave the overheating Earth.

HELIOCENE
The future time period or epoch when humans are no longer on earth, but are permanently and freely traveling within the solar system. From the word "Heliocentric," meaning centered around the sun, as in the solar system.

GALACTIOCENE
The future time period or epoch when humans are forced out of the solar system permanently and are traveling through the Milky Way galaxy.

COSMOLIOCENE
The future time period or epoch when humans are traveling the cosmos from galaxy to galaxy.

SCCT Single-Cell Computer Transporter
Single-celled animals that can propel themselves and have a memory cell or computational behavior.

MCCT Multicellular Computer Transporter
Multicellular life forms that propel themselves and have a memory cell or computational behavior.

ACT
Brain computational animals that propel themselves in water.

QCT Quadruped Computer Transporter
Brain computational land animals propelled by four legs.

BCT Bipedal Computer Transporter
Brain computational land animals propelled by two legs.

GCT Galactic Computer Transporter
Brain computational animals in space transported by a spaceship throughout the galaxy.

PETRI PLANET
The scientific description of a planet that harbors biology like a petri dish.

PETRI DISC
A solar system with a planet that harbors organic life using other orbital objects and or planets to support that life in the shape of a disc.

TCM Tin Can Man
A man that lives in a Tin Can or Space Ship.
A final product of the entire organic fungal life process, as a minimum example of the success of that organic *fungus* life or, metaphorically, a spore.

LE Lightless Epoch
The time or period when all the light in the universe is no longer generated.

MCG Maximum Conscious-Growth
When consciousness has grown to its maximum, when time has ended, when the universe has totally expanded.

ISE Inevitable Star Explosion
The fact that all stars eventually explode.

PPL Perspective and Perception Limitation
A means of generating an individual from birth with a view from a specific direction in time, space and a particularly evolved conscious state.

IPT In Present Time
A time when an entity is present within its present space and time.

PCG Pinnacle of conscious growth
The line between dark and light on the Earth or any biosphere (the terminator line) represents MCG of the *fungus* at any given time.

PAAF Plant and Animal Fungus
The original Earth fungus that began as both plant and animal equally.

CHAPTERS

1. COSMOLOGY

Introduction

The universe motivates consciousness to evolve by placing limitations on the ability for that consciousness to gain knowledge of its origins, the span of time in which an organism can survive, and the environment in which that consciousness can survive. These limitations are inherent in the universe's structure and are also the process guide rails by which consciousness is propelled to evolve through time. That is to say that these limitations are a universal constant and will not change as the universe continues to evolve and expand, which perpetuates the growth of consciousness throughout its vast expanses. These limitations are briefly described below:

TL;DR: The universe appears to have certain characteristics that continually impose stressors upon conscious-growths that guide them to grow in both consciousness and awareness. For example: I am conscious that I am not aware. I am aware that I am not conscious. These characteristics are woven into the fabric and physics of the universe itself. However, the three most fundamental characteristics are: limited information about the origin of all things, limited time for survival, and the limited ability to survive outside a very tight range of environmental conditions. These limitations are so profound that all life is subject to them and they provide constant stimulus, both physical and, later, psychological. These limitations also operate as a system that continually promotes the formation of life throughout the cosmos as well.

First Fundamental Limitation – Hidden True Origin (HTO)

Refers to the fact that any conscious-growth has limited knowledge of origin/creation by the very nature/structure of the universe in a way that, up to our current understanding, is not discernible where any conscious-growth originated or where it is going and why it is here (e.g. the Big Bang (BB) makes it difficult to discern a true starting point of the universe).

Second Fundamental Limitation – Inevitable Death

Describes that any organic construct or life form will meet its demise due to aging, predator/prey or the ever-increasing hostility of its supporting biosphere/planetary or post-terrestrial system. This Limitation is a giveaway to the function and purpose of the universe. Death is necessary for the advancement of conscious-growth to occur by constantly threatening the organic life to become inspired to live and adapt to an ever-changing environment. Inevitable death also leaves an example and insight to motivate the new organic life. And it is essential for the process that supports HTO and MCG.

Third Fundamental Limitation – Biospheric Bound

Dictates that any living entity must have evolved in a biosphere, starting from simple single-celled life forms and going on to evolve into more complex life. The complex life that evolves from this single-celled life must do so from the elements and raw materials found on its supporting biosphere. As a result, any excursion away from this biosphere would require the organism to bring an artificial version of its biosphere along with it (spaceship); it is still bound to the conditions of its biosphere (not biospheric-free).

Now that the mechanisms by which the universe promotes conscious-growth have been formulated, it is instructive to understand the timescales over which they operate.

THE EXPANSION OF MATTER
THAT SPAWNS AND PROPELS THE GROWTH OF CONSCIOUSNESS PART 1

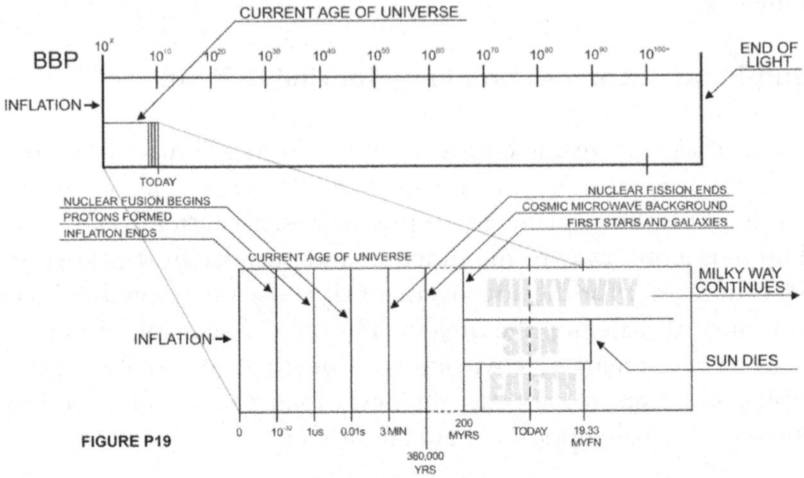

Figure P19 (The Expansion of Matter)

This figure shows the time in which matter has been expanding and the production of stars, galaxies, solar systems, and planets in order to grow a consciousness to maturity.

THE EXPANSION OF MATTER
THAT SPAWNS AND PROPELS THE GROWTH OF CONSCIOUSNESS PART 2

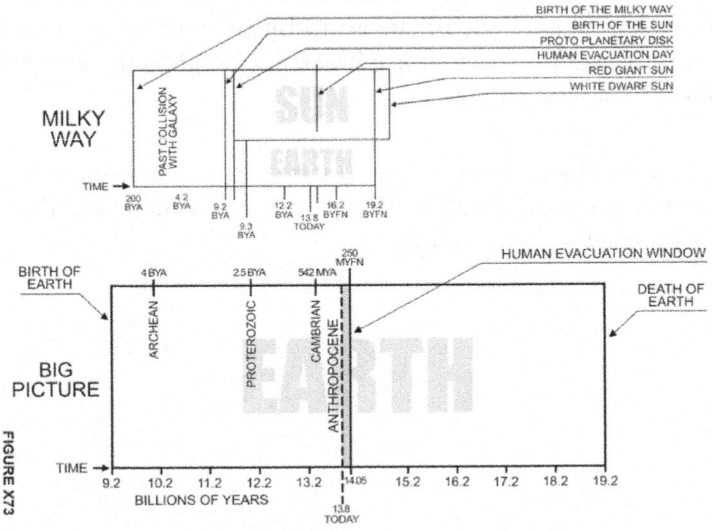

Figure X73 (Expansion of Matter Part 2)

This figure shows the time and function of the formation of matter that spawns a *fungus* to its maturity and out into space.

It will be important to present some form of evidence that these three limitations are actually present in the universe, which will help solidify an understanding and appreciation for Fungus Theory. In that vein, evidence that each of these limitations exists is provided and a description of how this evidence supports its particular limitation will be given.

Evidence of and Reason for Hidden True Origin (HTO)

TL;DR: The evidence for limitations on discerning the origins of life and the universe is apparent in the evolving structure of the universe and life on Earth. Factors like black holes, cosmic inflation, redshift, and Cosmic Microwave Background Radiation contribute to the challenges of obtaining information about the universe. The trend of increasing consciousness on Earth is accompanied by a growing number of unknowns and questions. Quantum mechanics, the Double Slit Experiment, and the inherent "fuzziness" of the universe suggest a Hidden True Origin (HTO) at all scales, obstructing the complete understanding of reality. As the universe expands and certain processes change, information will become increasingly hidden, leading to a state of burnout where the True Origin is concealed. Hiding the True Origin is seen as a mechanism to drive conscious growth by maintaining the pursuit of discovery.

Evidence for the fact that we exist in an environment that limits the possibility for discerning the origins of life and the creation of the universe can be seen in the structure of the evolving universe itself, as well as through observing the life that has evolved here on Earth for billions of years.

Black Holes

These structures act as the ultimate attractors in the universe. Even light is unable to escape their gravity if the path of the light beam is directed too close to the black hole. However, black holes allow for the formation of star systems within the matter that orbits them, which we call galaxies. Given their "dark" nature and the fact that they are powerful attractors, their role of being the progenitors of star systems is not obvious.

Expanding Matter and Cosmic Inflation

With the use of powerful telescopes, mankind has been able to determine that all galaxies are flying away from each other due to the fact that space itself is expanding. This is to say that the fabric of space, not the galaxies themselves, is expanding. This means that every galaxy (if after coalescing) is getting further and further away from one another throughout the universe. The galaxies themselves are not traveling through space but the space they occupy, relative to one another, continues to expand. It is this expansion that will ultimately separate galaxies from one another and then, ultimately, star systems such that information about any location in the universe seems unattainable as its light will never reach any other location that is not within that star system.

Red Shift

While humanity will not have to worry about its isolation from other galaxies due to cosmic inflation occurring in the near future, the universe is moving towards this state and the evidence is clearly visible by measuring the redshifted light from various galaxies. Redshift occurs when objects are moving away from an observer, like the sound of a siren of a fire truck after it has passed you on the street. The sound begins to decrease in pitch, due to the Doppler Effect. The same is true for light. The light from stars in other galaxies outside of our own seems to be redshifted to greater extents the further away it is from us. This would suggest that the universe is not only expanding but that this expansion is occurring at an accelerated rate the more distant a celestial body resides. This imposes obvious limitations on both interaction and information about the current state of the universe, as well as what has happened in the distant past.

Cosmic Microwave Background Radiation (CMB)

One might ask how science can know that this inflation came from some beginning event, as opposed to it being an ever-present state of the universe and that it always will be this way. Such a massive inflationary event would require a great surge of energy and this energy would leave an imprint on the cosmos in some way. It turns out that there is a dull hiss in the universe throughout its entire expanse. Everywhere scientists look, there is a small amount of microwave radiation, with minor fluctuations that have only been explored in detail due to recent advances in technology. This background radiation fits all models of inflationary theory, which would suggest such an event should leave radiation of this form lingering in the background of the vacuum. This background radiation is called the Cosmic Microwave Background (CMB). Thus, the evidence suggests that the universe is operating via a mechanism that intrinsically obscures information from observers through continued expansion.

Evolving Consciousness Itself

When viewing the various forms of consciousness that have evolved on Earth, there is a definite trend toward increasing awareness, mental capacity and technical ability. However, as the level of consciousness has increased with time, the number of unknowns and questions has only increased. Humans today realize they know far less about the universe than they did in the mid-1900s when the standard model was supposed to be the final frontier of understanding physics and the various properties of matter and the processes it undergoes. However, with the discovery of the Higgs Boson and other possible particle discoveries around the corner, humanity is once again humbled by the amount of information it has yet to uncover, as well as the questions about reality that these discoveries bring. Most notably, our understanding of quantum mechanics and its ramifications for things like consciousness and the very fabric of the universe itself suggests that Hidden True Origin (HTO) is the most fundamental aspect of reality itself. Inherently, the universe is "fuzzy", as was demonstrated by Heisenberg's Uncertainty Principle. It is this inherent "fuzziness" that demonstrates the universe has an element of uncertainty in it, which makes it nearly impossible to trace back a True Origin (TO) to all life and the universe itself. One might think that this is insufficient to describe something as large as the universe; however, the universe started in a quantum state before matter condensed and began following behaviors at the macroscopic scale that we deem as not being tied to a probabilistic outcome but a determinate one.

Furthermore, the modified "Double Slit Experiment" demonstrates that due to the fact that when one tries to determine which slit a single photon goes through when fired at a two-slit screen apparatus, the standard wavelike behavior that is normally observed when one is not trying to determine which slit the single photon goes through is destroyed by any attempt to make a such a determination; particle behavior returns. This is further evidence that the universe is hiding information in the form of collapsing the totality of possibilities that are contained within the wave function of that particle when a "measurement" is made. This blurs reality as what is happening when no measurement is made is vastly different from when a measurement is being made. Physicists still are not sure why or how this process even occurs or why it must occur this way. However, what this does demonstrate is that if consciousness is built from the smallest aspects of reality upward, at each level there is a system in place that hides information from a growing consciousness. Whether this is at the scale of large structures expanding with the universe, complex genealogical and archeological records, or the very fabric of what makes up the matter of the universe, information is hidden in some form. Therefore, Hidden True Origin (HTO) occurs at all scales and evolving consciousness is subject to the limitations it imposes. Thus, Hidden True Origin (HTO) is Evidence of System (EOS) that is universal and is required for growing consciousness and consciousness would stop growing without them.

Burnout

As the universe continues in its accelerated expansion, some processes will decrease in frequency, such as star formation and others will increase in frequency, such as black hole mergers and the devouring of galaxies by their central supermassive black holes. When the universe has expanded to such a size that the light from other galaxies is imperceivable due to the expansion occurring at speeds faster than light and, subsequently, all stars have died, information will only exist in the form of gravitational waves and electromagnetic fields outside the visible spectrum. After this point, it will be the case that black holes engulf all the matter in the universe, radiating away their energy in the form of Hawking Radiation. At this point, the universe will have gone through an erasure, completely hiding any information about the origins of life and the existence of matter that permeated it. Once again, the True Origin (TO) of the universe and all life within will have been concealed.

The reason behind hiding a true origin from a conscious-growth is: when a conscious-growth becomes conscious enough to recognize that it has no way of obtaining the information of its True Origin, it will fight to find this discovery for as long as it is withheld. If a conscious-growth were to obtain this information, it would no longer search or have the need to act without the need to find, making action no longer an important drive. Therefore HTO is the main reason for the continuation of conscious-growth to its MCG.

Evidence of and Reason for Inevitable Death

This may feel quite obvious to most individuals as we are all faced with our mortality on a daily basis. However, it is important to gain insight into the destiny of various structures throughout the cosmos and how this affects the habitability of any one particular region of the cosmos throughout all time.

Star Lifetimes

The fact that the Earth and other planetary biospheres, are at the mercy of the star about which they orbit, having temporary periods in which they are hospitable to life, demonstrates this limitation. Not only are stars volatile on cosmic timelines, but there is also limited time for which the universe can support star formation. Thus, wherever life may form and is bolstered by a star, that life is ultimately propelled outward by its star's increasing output in order to survive. If that life should take root in another star system, the result will inevitably be the same. Thus, the universe ultimately forces life to evolve to the point of interstellar sustainability in the vast reaches of the cosmos, unbound to any star system, through inevitable death.

Black Holes

As mentioned before, these interstellar attractors are responsible for the development of increasingly complex structures of matter, which mainly consists of galaxies that contain stars and gaseous formations. However, they are simultaneously extremely dangerous, which acts as a constant stressor for any conscious-growth that should form near the galactic center. These conscious-growths are the ones that will be threatened with death the soonest by being sucked into the black hole. Over long periods of time, entire galaxies will succumb to the extreme gravitational forces of their host black holes.

Burnout II

As mentioned before, the universe will reach a point where all matter has been engulfed by black holes and re-radiated in the form of Hawking Radiation, which will have erased any sense of True Origin

(TO). However, this also readily demonstrates that the existence of any formation of matter is temporary. In order to survive within the universe, adaptation and mobilization in order to escape these temporary structures are required.

Evolving Consciousness II

Everything organic that exists today did not originate in the physical form it currently possesses. Life has mutated and evolved from the passing of genes from one generation to another in an attempt to escape inevitable death, which has been the system responsible for this sequential development and evolution of conscious-growths, from early single-celled organisms up to the bipedal human organism today. This is Evidence of System (EOS) for the growth of a Collective Consciousness (CC) that is recognizable today as every cat, pollywog, and human. While this sounds far-fetched, it is actually much more sensible than you might believe. Humanity has been proven to be a collective consciousness that is capable of reacting to world events on a global scale in a measurable way. The Global Consciousness Project (GCP) created originally in the Princeton Engineering Anomalies Research Lab at Princeton University, is directed by Roger Nelson from his home office in Princeton. The Institute of Noetic Sciences provides a logistical home for the GCP.

"When human consciousness becomes coherent, the behavior of random systems may change. Random Number Generators (RNGs) based on quantum tunneling produce completely unpredictable sequences of zeroes and ones. But when a great event synchronizes the feelings of millions of people, our network of RNGs becomes subtly structured. We calculate one in a trillion odds that the effect is due to chance. The evidence suggests an emerging noosphere or the unifying field of consciousness described by sages in all cultures."[1]

This project has been measuring the deviation of random number generators from their nominal random output in correlation with impactful world events in which consciousness is polarized. What they have found over 30+ years of operation is that the entirety of humanity is capable of reacting to global events and, even more interestingly, their reactions to these events occur in a way that is anticipatory, occurring prior to the actual event by up to a few hours.

The need for inevitable death is to cultivate conscious-growth through a baton process. Each individual conscious entity's perspective is spent and recorded through its actions leaving behind the pros and cons of those actions for the next generation to propel forward towards TES. It is used and translated as information and story.

Biospheric Bound

Evolving Consciousness III

The fact that most organisms can only survive in habitats that closely resemble their supporting biosphere and not in dramatically different environments (such as outer space or other biospheres) is evidence that the universe places limitations on the range of variability in organisms' habitats. Species go extinct as these tolerance bounds are exceeded, which has been observed through various fossil records as well as more recently as species are going extinct at an alarming rate due to global changes in climate that occur cyclically (e.g. Ice Ages, etc.). Some scientists argue that we have already entered the next mass extinction. This demonstrates that life must be bound to a biosphere which is the third and final fundamental limitation used in this theory.

By obscuring information that would disclose an origin to life in the universe and forcing organisms to exist on temporary biospheres in which they will die or evolve the means to escape their death, the three fundamental limitations work together to spawn and propel the growth of consciousness throughout the universe. We humans are the greatest example and most obvious evidence of the result of these limitations to date.

The reasoning behind this limitation is to force or enforce the fungal collective growth of that biosphere to act with responsibility without escaping or avoiding, this will enforce a greater conscious awareness. This true and reliable progression can only be done by never escaping the task in front of the entity. The end of this biospheric life is also a forced motivator along with hidden True Origin that is the cause of all actions made by a conscious-growth. There would be no accumulation of conscious-growth without this limitation, technically or emotionally.

How the Three Fundamental Limitations Spawn and Guide Conscious-Growth

TL;DR: The concept explores three fundamental limitations that shape the growth of consciousness, creating a void known as the "Void Of Psychosis" (VOP). This VOP compels conscious entities, particularly bipedal organisms, to evolve and strive for a greater understanding of their existence. The limitations enforce a terrestrial confinement, driving the need for survival and evolution. Evolution involves technological advancements and the sharing of knowledge through cycles of life and death. The overarching purpose is to achieve Maximum Conscious-Growth (MCG), potentially leading to a Uni-Galactic Conscious Harvest (UGCH) and continuous exploration beyond galaxies. The narrative outlines the intricate process of consciousness growth, involving the expansion of the universe, the formation of star systems, temperature-controlled biological evolution, and the development of consciousness. Ultimately, the goal is to reach a state of Maximum Conscious-Growth in a technological biosphere.

The three limitations work in conjunction with one another to distract those subjected to them from detecting that they are at work. They are to be generally accepted as absolute conditions because of the fact that they demonstrate themselves as evident throughout every observation. As life begins to take root on a biosphere, various mechanisms of adaptation begin to express themselves when stressors are imparted upon that life. Once a consciousness of sufficient complexity is formed through the continual adaptations that have taken place in order to promote continued survival, observations give way to a realization that the True Origin (TO) of life is not known. This happens in an observable universe that hides that conscious awareness from this TO and as a result, "a void" is formed in the entity's personal identity. This void develops into a kind of subclinical psychosis within an entity, or Collective Consciousness (CC) of many entities, that causes a reactionary response to compensate for this lack of knowledge. This condition can be termed the "Void Of Psychosis" (VOP). This void forces that conscious entity to engage in actions that will form a more concrete identity.

It is important to note that the presence of a Void of Psychosis (VOP) occurs only when this Collective Consciousness (CC) evolves to a point where bipedal organisms have evolved. Thus, this Void of Psychosis (VOP) is a vehicle of both a physiological and a psychological evolutionary growth process. This Collective Consciousness (CC) is allowed to increase in consciousness and awareness only to the extent that it is permitted by the guiding principles and physics of the simultaneous enforcement of the three fundamental limitations. These operating dynamics fragment and distort any determination of both the purpose of life's existence, as imparted by knowledge of its TO, and the TO itself. These conditions promote the evolutionary growth of that Collective Consciousness (CC) by forcing it to strive for greater understanding and bolster survival through the development of a more complete identity. A more in-depth discussion on the Void of Psychosis (VOP) is included in Chapter 5.

Being conscious and confined to a terrestrial biosphere is a forced condition due to the third limitation, which makes it impossible for any Collective Consciousness (CC) to escape its worldly task of constantly struggling to escape its demise (enforced by the second fundamental limitation). The third fundamental limitation forces any Collective Consciousness (CC) to work through all terrestrial encounters, ultimately for its growth and development, so that a successful terrestrial evacuation is ensured. This is required as that Collective Consciousness (CC) will not survive beyond the end of its planet's habitable period unless the evolution of a new kingdom (the proposed Eighth Kingdom of technology, discussed later in the chapter) occurs. However, the evolution of The Eighth Kingdom alone is not enough. It must become sufficiently developed in order to bring its biosphere and the technologically advanced bipedal organism with it beyond the confines of the planet. As a result, it forces the intelligent evolution of the CC.

To ensure that technology and information are shared ubiquitously so that the entire Collective Consciousness (CC) can benefit from the experiences and growth of each individual, continual cycles of life and death must occur. Due to the ramifications of death, conscious organisms will be pressured, in some sense, to ensure that their work and knowledge are not lost forever. They will leave recordings, teachings, stories and information behind, handing the baton over to a new generation for the task of expanding knowledge and the continual process of innovating that is needed to survive the end of their planet. Without inevitable death, an emotionally intelligent organism's consciousness would become stagnant with no need to innovate.

Overview of How Consciousness Is Grown

This is done with a hands-off method of igniting consciousness, growing more conscious without knowing its place of existence before its known consciousness now. Simply erase any observable point or origin of matter and its ignition and it is forever undetectable.

1. First, the universe begins expanding from some centrally located region. Any determination of this location would need to be done far in the future by sufficiently intelligent conscious entities, however, by this point in time, the universe will have expanded so greatly that any determination of the origin point of expansion will be impossible (the first fundamental limitation). This is a method that prevents the traceability of what grows consciousnesses that are aware enough to observe themselves in the environment in which they reside (the universe).

2. As the universe continues to expand, it also cools. This process allows for the clustering of matter and the formation of large-scale structures. These structures include black holes, stars and galaxies. Initially, the stars either burn out after some lengthy period of time and go nova, become dwarfs with extended periods of slow death, or they turn into black holes. In the case of stars going nova, after having fused elements of their entire lifetime, these exploding stars seed the entire universe with the material that is required for more complex structures such as planets, asteroids and meteorites. Black holes, on the other hand, will continue to attract material and grow in size, eventually reaching sizes so large that they can host entire collections of matter around them hundreds of thousands of light years across. These structures are the galaxies capable of birthing stars at higher rates due to the fact that gasses and other materials for star formation are more centrally located as they spiral inward toward a black hole. Once these structures have formed and stars have spread their newly generated elements throughout the cosmos, stars can now act as progenitors of organic life on the biospheres that form around them in protoplanetary disks. These protoplanetary disks are now capable of housing a large variety of materials when compared to the disks of hydrogen and helium that would have formed around the very first stars. This changing nature of stars' role in collecting matter around them in their orbits is in alignment with the principle of Hidden True Origin (HTO). This is the case in that their behavior at the time when a conscious

observer could view them would be different from their behavior in the early universe. Once again, information is obscured with time and keeps consciousness guessing at any purpose for their existence.

3. Meanwhile, evolution on the planetary scale is a process that also enforces HTO and operates through dynamic thermal regulation over billions of years. Initially, the planets cool from their extremely hot protoplanetary form. This allows life to begin to form in its primitive state, starting as a single-celled organism. Subsequently, the planet begins to reheat due to the nature of increasing star output with time. In the interim, life adapts to its environment and grows increasingly conscious of its environment and situation. Over time, more and more fossil records are lost due to erosion or other destructive geological processes. This makes it nearly impossible for conscious-growths (when they become capable) to link together enough evidence to trace the evolution of all species completely or discern their exact origin (Darwinism). This makes the full traceability of the evolution of all life something the Collective Consciousness (CC) cannot be cognizant of as an absolute and is thus another way in which HTO is promoted, which enforces a continued growth and expansion of consciousness. Like the erosion of fossils, the expansion of the universe leaves only traces of evidence in the background (CMB) radiation persistent in the universe today. We can see that terrestrial evolution has a kind of built-in system that promotes and ensures HTO on all scales.

4. After the development of organic life, any stress on these biological systems will promote continued evolution in the form of adaptation. The second fundamental limitation is enforced through a variety of stressors, such as starvation, which enforce predator-prey relationships and ultimately propels the foundation of brain growth. Through continual adaptation with multiple species competing, this generates a need for more complexity in consciousness. This increase in complexity allows that consciousness to survive for many generations and represents the growth process by which continued evolution occurs, pushing CCs toward MCG. On the path toward achieving MCG, a Void of Psychosis (VOP) eventually forms within the psyche of sufficiently evolved organisms that acts to further propel conscious-growth. This is how a brain capable of creating the technologies required to leave its birthplace evolves inside a bipedal organism (BCT).

5. Once these technologies have been created, the BCT (Bipedal

Computer Transporter) will begin to be forced out into its star system due to the increasingly hostile conditions of an inevitably heating star. This BCT will be forced to exist as a Tin Can Man (TCM) as it escapes the hostility of its host star.

Eventually, venturing out into the star system will be insufficient as the threat posed by the black hole at the center of a Collective Consciousness' (CC's) host galaxy forces that Collective Consciousness (CC) deeper into interstellar space. The end product for all organisms in this stage is a GCT (Galactic Computer Transporter). Throughout each galaxy, of which there are an immense number in the universe, an even larger number of biospheres organically grow consciousness independently. This independent growth allows for the optimal evolution of a species such that it may survive past the end of its organic birthplace. These conscious-growths will likely interact with one another at some stage, which promotes the continued growth of consciousness through the sharing of technology and culture. Ultimately, the universe continues to enforce conscious-growth at every stage of its development. These conscious-growths will push one another toward MCG and prepare one another for the Uni-Galactic Conscious Harvest (UGCH).

BP Over time, as all life on Earth reaches its expiration date, man will hopefully evacuate this biosphere and embark on a journey outside the solar system to avoid annihilation. In doing so, it will be forced to become a Tin Can Man (TCM), at least one technologically advanced being in a spacecraft that supports eternal self-sufficiency (ESS).

BBP Eventually, in whatever form a matured conscious entity will have evolved to by this point, fleeing the galaxy will also become necessary as the black hole at the center of their galaxy will consume the stars that surround it. Ultimately, this is a continuous design of escaping inevitable death and searching for TO. The continuous search for TO, until the point in which light is either no longer emitted, as all matter has been swallowed by black holes, or the rapid expansion of the universe leaves each local region in total darkness, the Lightless Epoch (LE) would render a Maximum Conscious-Growth (MCG).

The only thing left would be organic consciousness inside a technological biosphere.

BB + LE = MCG!

EOS

- Formation of star and solar systems HTO
- Temperature controlled biological growth CC
- Development of a VOP
- Growth of a consciousness

Physical Evidence that the Universe Is a Structure for Conscious-Growth & How Consciousness Is Grown

TL;DR: The universe serves as a structure for conscious growth, with physics indicating that all matter is generated to support three fundamental limitations, propelling consciousness to acknowledge its own mortality. A changing environment, driven by the random development and dissolution of matter, promotes conscious growth throughout the cosmos. Supermassive black holes, found at galaxy centers, obscure information about True Origin (TO) and life's origin, fostering a diverse and hidden mechanism for consciousness formation. Stars' activity, planetary formation, and cosmic events act as stressors, essential for evolution and technological advancement. These events, though potentially destructive, are crucial in shaping ecosystems and achieving Technological Evacuation Standards (TES). Stars' limited lifespans enforce the second fundamental limitation of inevitable death, ensuring conscious growth by pushing life to survive and evolve. The timeline from the Big Bang to Burnout dictates the schedule for conscious growth, forcing organisms to evolve, become aware of their situation, and develop technologies for space travel. The universe, through its construction method, hides True Origin, encouraging consciousness to grow and evolve while avoiding stagnation.

The random development and dissolution of all matter, either through the extremely rapid expansion of the universe or the immense gravitational pull of supermassive black holes, is responsible for propelling conscious-growth throughout the cosmos. These supermassive black holes are found at the center of what appears to be every galaxy. A majority of stars are bound to them as they orbit around their host black hole. This structure is maintained for billions of years, though many stars are ejected from their host black hole in the collisions that can occur between galaxies or in mergers between galaxies. Since these rogue star systems are much harder to detect, due to their limited light output in contrast to the background of galaxies that surround them, they are elusive. This behavior, and thus function, of supermassive black holes hides information about both TO and the origin of life in any galaxy, making it harder to detect the process by which the formation of life in star systems is carried out. The very fabric of the universe itself- a cold, dark and "soundless" vacuum as an environment- provides support for the nuclear fusion of elements (star formation) en masse, for extended periods of time. Stars, like our Sun, act as the heating element within their systems which, in turn, drive the random aggregative processes that form planets, as well as bombardment processes that deliver various minerals and compounds to these planets. This fosters the development of proto-biospheres, a rare few of which will end up developing organic life due to their proximity to the star being in a temperature zone that allows the formation of liquid water and other organic precursors. It is on these planets that the conditions allow for conscious-growth to begin. The fact that some planets can exist in rogue systems and others bound to galaxies, the random aggregative process of star formation and the temperature requirements for the formation of organic substances all act to obscure the mechanism by which life is formed throughout the cosmos. These are necessary components to hide the intended structure for the formation and growth of consciousness throughout the universe.

For every star, there is a region that lies at a range of distances which would allow the proper thermal energy transfer in order to develop a biosphere from a planet that can sustain liquid water and other organic compounds. Larger stars would have this range of distances existing further away; and in smaller stars, the opposite would be true. Since heavier elements and compounds are found on planets closer to their stars and lighter elements are found on the outer planets, there exists some level of predictability to what can be found on inner planets versus outer planets. The true determining factor is whether water will exist in conjunction with these other elements and compounds. Ultimately, everything organic is temperature-controlled—just add "water."

In the early development of the solar system, there was only gas and particulate matter circling the sun like a large flat disk. This disk of debris, called a protoplanetary disk, starts a process of gravitational attraction, where smaller arrangements of matter collide and form larger arrangements, which results in a coalescing of matter through a continued process of bombardment. This increases the gravitational pull of some regions within the star system where the mass has accumulated to create larger bodies. Ultimately, it results in planet formation. In our particular system, and as is common in most systems, the hard rock planets form nearest to the star due to the fact that heavier elements are drawn further inward. These planets are exposed to much higher levels of radiation and heat, which causes the material on these planets to initially melt and reform after some time of cooling. All other matter further from the star, or past the asteroid belt as in our system, accumulated as gas planets due to lower levels of heat/solar radiation and the lower mass of these elements. Ultimately, all planets came from some mixture of gas and solid matter in varying ratios.

Star activity is not constant. Stars get hotter as they age and they spew out massive amounts of energy at times as they flare. This can stunt the growth of biospheres and, in a star system with a powerful enough star, potentially wipe out all life on a biosphere. Thus, the universe is always imparting stress in one form or another on biospheres that form, in accordance with the second fundamental limitation.

The fact that the universe is not constant and is expanding also gives each of these stars protection against the others through relative isolation. This is critical as the radiation from one star has very little effect on the others and, a majority of the time all that can be perceived is a twinkle in the night sky when observed from any given biosphere. However, every so often, stars from other star systems can influence the development of biospheres as they go nova. This sudden burst of radiation can possibly cause temporary extinctions, or radically shift the ecology on a biosphere should its light reach a planet during a period of time when various forms of life are thriving. As evidence of the need for stars to generally exist outside the region of influence of other stars, scientists believe a mass extinction event can be linked to a gamma-ray burst in Earth's distant past. The gamma-ray burst was so powerful that it wiped out a significant portion of life on Earth. A supernova may have been responsible for this gamma-ray and the subsequent mass extinction in Earth's history.

"Toward the end of the Devonian period (416 million to 358 million years ago) there was a mass extinction known as the Hangenberg Event; it wiped out armored fish called placoderms and killed off approximately 70% of Earth's invertebrate species. But scientists have long puzzled over what caused the die-off. Recently,

preserved plant spores offered clues about this ancient extinction. Fossil spores spanning thousands of years at the boundary of the Devonian and the Carboniferous periods showed signs of damage by ultraviolet (UV) light. This find suggested that a cataclysmic event had caused a long-lasting disruption of Earth's ozone layer, which shields the planet from harmful UV rays. Scientists proposed that a likely candidate for this blast of UV light could be one or more supernovas that exploded within 65 light-years from Earth, according to a new study."[2]

This acted as a stressor to Earth's ecosystem in order to select the fittest creatures. While this event could seem like a doomsday event, life rebounded in a robust fashion on Earth and modern humans are the result of that later explosion of life that occurred thereafter.

While these events seem both violent and destructive, with the potential to wipe out all life in a biosphere, they are likely essential on the path for any ecosystem of organisms to evolve to meet the Technological Evacuation Standards (TES) of that biosphere. Every planet has slightly different required technological standards due to its mass & gravity, atmospheric density and atmospheric pressure systems. Thus, each conscious-growth will have a slightly different path toward evacuating its biosphere; however, the overall path is similar enough in its final goal of achieving Eternal Self-Sustainability (ESS) that the ultimate lessons learned and technologies developed will be quite similar.

Since stars have a limited lifespan, they also help ensure conscious-growth by imposing the second fundamental limitation of inevitable death. These stars not only reinforce the life that forms but also force it to endure and survive its inevitable demise, in turn bolstering conscious-growth.

In order to understand how consciousness is grown across the universe's planet-ation, one needs to acknowledge the scope and timeline through which it appears. With the increasing expansion of the universe and the creation of new elements through fusion within stars, the formation of stars and their star systems, and the mixing of matter by way of ejections in supernova and random comet/asteroid impacts, organic life begins to form through these cosmic interactions. This organic life will continue to evolve as it struggles to survive on an increasingly hostile biosphere. In order to do so, it must become aware of its situation within both its ecosystem and, more importantly, the universe. This promotes consciousness and, thus, conscious-growth. This timeline (Bang to Burnout) dictates a schedule that forces the evolution of these conscious-growths that possess technologies by requiring that they survive the end of their birth planet as they are pushed out of that nest into outer space. The process of selecting any such organism that surpasses the need to be bound to its biosphere demonstrates that conscious intelligence is farmed by the universe. After a certain level of evolution in that conscious-growth, it becomes clear to that consciousness that while it may not know its true origins, it does now perceive that its true origin is hidden. This process unfolds in many steps over eons and eons.

If true origin were found, being cognizant of it could render a consciousness under motivated so as not to meet its biosphere's technological evacuation standards (TES). Random development in these systems is also evidence of a fundamental process that helps ensure the ignition of life, if possible, anywhere in the expanding universe. The formation of ordered large structures-such as planets, star systems and galaxies-from disordered distributions of smaller structures through the attraction and coalescing of matter appears to be a construction method that hides true origin in an elegant and efficient manner. This construction method was determined pre-bang, as the behaviors of the universe were determined at the earliest stages of cosmic inflation. Logically, this provides a great deal of support that our universe, being crafted as it has, has at its core the principle function of hiding True Origin from consciousness as that consciousness both grows to become aware of its environment and attempts to avoid inevitable death.

In order to reach a certain level of conscious awareness, the organisms must evolve to become bipedal. It is this form that maximizes the functionality of the organism and allows it to manipulate matter with its appendages in an articulate manner. These organisms are the ones capable of technological advancement.

EOS

- No traceable evidence of Original Origin
- No traceable evidence of a beginning
- No evidence of the a predictable future
- No sound travel in space
- The evidence of evolutionary process

- The growth of a consciousness
- A consciousness baked to maturity
- Forced off its biosphere
- Forced to develop technologies
- Forced biospheric evacuation
- Forced alternative biospheric existence

A Black Hole

TL;DR: By design, an environment that supports the existence of black holes is EOS (Evidence of System). The development of black holes is essential in birthing and nesting of stars and biospheres that are necessary for the success of a *fungus* system. This forcing process is done by subjecting that consciousness to growth with constantly adapting to its ever-changing environment. Think about it; if the growth of a consciousness depends on a long, slow growth process that involves the three fundamental limitations, there's no other obvious way to grow a consciousness. If the goal is to get consciousness to grow in an environment that forces it to adapt and evolve with no detectable beginning or end, this will force anything conscious to act. The only way to achieve this is in a cold, dark vacuum, filled with random, ever-developing Black Holes.

Black holes are an enigmatic formation in the universe. First theorized and then later indirectly observed, these structures push our current understanding of physics to the limit. Black holes can form when enough matter occupies a specific region of space such that the gravitational pull of that matter is strong enough to pull everything, including light, into its central region. These gravitational objects are the most powerful objects we currently are aware of and it is only recently that we are beginning to try and peer into the innards of these enigmas. Interestingly, there exist two well-known types of black holes: stellar-mass black holes and supermassive black holes. Stellar-mass black holes are formed by the collapse of large stars, though scientists are unsure exactly how supermassive black holes form. A third type of black hole, intermediate-mass black holes (IMBHs), has been theorized, and more recently some astronomers say they have detected the first true signs of these IMBHs. The prevailing theory is that these lower-mass black holes continually merge until they grow to reach the supermassive scale. More research is needed in order to determine exactly the dynamics of how black holes can grow to such gigantic proportions. However, we can clearly see that they do exist.

Recently, scientists have proposed that there is evidence that the Milky Way Galaxy merged with another galaxy some 10 billion years ago. By analyzing the orbital dynamics of various stars within specific regions of our galaxy, researchers came to the conclusion that the behavior of a certain group of stars near the center of our galaxy was moving differently than the rest, which was also in accordance with computer simulations of galactic mergers.

"The halo of the Milky Way mainly comprises the remnants of a pre-formed structure of stars that merged into our galaxy about 10 billion years ago, researchers in France and the Netherlands have concluded. The researchers studied new data from ESA's Gaia Space Observatory alongside spectral information from the Sloan Digital Sky Survey, and they concluded that stars from the object, which they have termed Gaia-Enceladus, are chemically and kinematically distinct from other stars that formed from dust and gas in our galaxy."[8]

If galaxies are indeed merging, and ours has already, it is no stretch of the imagination to suggest that smaller stellar-mass black holes can merge to form IMBHs and IMBHs can merge to form supermassive black holes. Scientists have also discovered that our galaxy, Milky Way, is on course to merge with our nearest neighboring galaxy, Andromeda, in the distant future as well to form "Milkdromeda."

If these mergers continually produce larger and larger black holes, then immense amounts of gravity can be housed within a localized region of space. Scientists have also been theorizing about another type of black hole, called a "primordial black hole," which formed in the early universe much before star formation and subsequent star death could generate the first stellar-mass black holes. These black holes would also be very large in size due to the nature of forming during the early inflation of the universe. While scientists aren't sure of their existence conclusively, there does seem to be a lot of evidence for them.

"Then there are primordial black holes, much more mysterious objects that are thought to have formed soon after the Big Bang. The thinking is that random fluctuations in the distribution of mass in the early universe must have created some regions dense enough to form black holes. However, nobody knows if primordial black holes actually exist. Astronomers just haven't been able to gather the evidence. That is beginning to change. In 2016, astronomers began operating a gravitational wave detector called LIGO that can measure the way the universe rumbles when two distant black holes collide. Since then, they have spotted 47 collisions between black

holes of all kinds of different masses. And that has given them an interesting database to study. Now the latest analysis suggests that more than a quarter of these collisions involved primordial black holes. 'If confirmed, these results would imply that the LIGO/Virgo collaboration could have already detected up to 24 binary black holes formed in the early universe,' says Gabriele Franciolini at the University of Geneva and colleagues, who carried out the data analysis. This would be the first observation of any primordial black hole."[4]

Regardless of whether a supermassive black hole hosts a galaxy or a primordial black hole does, both structures possess the immense gravity required to sustain and support all the material that exists within a galaxy. Supermassive black holes are found at the center of almost every large galaxy (ours is called Sagittarius A). However, black holes can also be found randomly scattered throughout the universe, without an orbiting galaxy, but they are much harder to detect due to the fact that they aren't feeding on any matter or bending the light of the galaxy they host around them. In the case where they have attracted large amounts of matter, they act as gravitational axes about which entire galaxies rotate. These cosmic attractors act as stellar nurseries forming nebulae, stars, and planetary systems all of which rotate about the black hole for billions of years. In this way, black holes act as a kind of womb that nurtures an embryo. These embryos are the biospheres that form in planetary systems of various stars in their large orbits around these black holes, for the eventual spawning of organic life. The orbits of these stars and star systems eventually become very stable, following predictable patterns that give these star systems the necessary environment required for the coalescing of various compounds and formation of structures that will promote the development of organic life.

Sagittarius A has a mass of four million solar masses (four million times the mass of our sun) and has kept our galaxy intact for many billions of years. However, in terms of what is common for galaxies, this is actually on the smaller side. There are some supermassive black holes at the centers of galaxies that are so massive they dwarf even Sagittarius A.

A gargantuan black hole, known as J2157, 34 billion times the mass of our sun mass and about 8,000 times bigger than the black hole in the center of the Milky Way, gorges on nearly the equivalent of one sun every day, according to Dr. Christopher Onken at The Australian

National University (ANU) about a monster of the early universe that brings an end to time and space and the laws of physics. If the Milky Way's black hole wanted to grow to that size, 'it would have to swallow two thirds of all the stars in our Galaxy,' he added.[5]

While supermassive black holes allow for stability in galaxies so that life and consciousness may form, they also can stress/threaten it too, which promotes accelerated growth. Given the ever-decreasing orbits, or in spiraling, of all matter in any orbit, the star systems nearest the center of the galaxy are the ones that will likely not possess or form conscious-growths. This is due to the fact that they are most likely to be gobbled up by their host supermassive black hole or decimated by intense X-rays that are ejected from the accretion disk of the supermassive black hole as it feeds on material passing too close to the event horizon. It is because of the inhospitality of this innermost region that life is most likely to form further away from the center of the galaxy. In fact, our Solar System exists on the outer two-thirds of one of the spiral arms of our galaxy.

Over time periods that are mainly unfathomable to humanity, many orders of magnitude longer than the time by which the last light from the last star is emitted, black holes will have all died a slow death due to a special process only recently discovered, called Hawking Radiation. Black holes will slowly lose energy by radiating away particles in a quantum process at or near the horizon of the black hole. Eventually this radiation will carry enough energy away from the black hole until it has fizzled out of existence. This fulfills their other role of being the cosmic erasers of the universe. They have, at this point, totally scrambled any information of the universe that existed before them. This enforces HTO in its ultimate form, as after this point very little can be gleaned about the origins of life and the process by which the universe unfolded over the many eons prior.

BBP Black holes' gravitational axes, like axles on a pinwheel firework, form randomly and act as a central focal point for matter to develop around as it is all slowly attracted to it throughout the millennia. This construct is highly efficient at blurring the reality from a consciousness capable of cosmic observation that this structure was intended from the start. Their different sizes, gravitational intensities, the timeline over which they develop, and the scattered placement throughout the universe compound the evidence for the concept of Hidden True

Origin (HTO). Black holes are seemingly enigmatic structures that baffle those that gaze upon them, both in their size and power, as well as purpose. However, they are the primary structures by which the universe allows consciousness to spawn and spring forth on planets throughout the cosmos. The extremely long time frames for which they can host matter around themselves pushes conscious-growth to its maximum (MCG) by protecting it and subsequently threatening it over very long timeframes. As long as black holes continue to ingest matter and radiate away in this fashion, their primary purpose will be to promote MCG by the mechanisms described here.

The Possibility of a Pre-Bang Life Dust and its Purpose

TL;DR: The Life Dust hypothesis suggests a pre-Big Bang ingredient or element, possibly present in the singularity, initiating conscious growth wherever conditions permit life post-Bang. Unlike panspermia requiring multiple random events, Life Dust ensures the Big Bang's success by being part of the expanding matter. Though atoms formed after the Bang, Life Dust might leave evidence of its distribution, providing insights into life's origin. It makes sense that if a creator were to cause a Big Bang, they would make sure (insurance) that the expanding matter, changing form and spreading out to form the universe, is the only random factor in this phenomenon. Otherwise, it seems there is a potential for the Big Bang to be a dud. Observing life on Earth hints at the potential existence of primordial Life Dust.

How the Life Dust Hypothesis Works with Fungus Theory

BP A petri planet is made possible only by the appropriate conditions set up by its petri disc (protoplanetary disk). With a planet at its appropriate orbital distance, the reduction in planetary bombardment (e.g. asteroids impacting the planet) will eventually allow planets to cool to a more suitable temperature for life (biospheres) to begin. The phrase "just add water" is ubiquitous for its denoting the fact that a process will readily occur merely by adding one simple element. Similarly, in the universe, the addition of water and/or various other ingredients will eventually lead to the formation of an original fungus. However, it is most likely for life to form under conditions that require less specific requirements to be satisfied in order for it to begin. Thus, temperature is the prime (only) candidate in this regard since temperature determines whether water exists on a planet. This single random variable determines the formation of a biosphere. If every galaxy had one Earth-like planet, a pattern would be revealed. Such a pattern would suggest the intent to grow consciousness in each star system. This realization would be against HTO, stunting conscious-growth. Therefore, the universe strictly employs a method of HTO for MCG by hiding the requirement for the development of a biosphere in the random variable of temperature so a system cannot be easily detected. This process repeats ad infinitum, acting like little lab experiments throughout the universe, seeing which ones will ultimately cultivate the form of life that can meet its TES.

BBP Life Dust may even be all dust (particulate matter and debris, and eventually minerals) in the sense that the requirement of more variables to match up in order to start life exponentially decreases the chance of that life beginning in the first place. If H2O at the perfect temperature is the only needed but rare element to start life, then it has the greatest chance to form at least once during a universal epoch. This strongly suggests that all dust is Life Dust.

By hiding Life Dust in plain sight and only requiring water for originating life, True Origin remains elusive. For if the evolution of the universe thus far did not make use of this ubiquitous Life Dust, what would life have had to rely on for its inevitability? Without it there could very well be a Big Bang that resulted in a dud, which implies that there likely is Life Dust of some kind.

Conscious-Growth through the Evolution of a Black Hole

TL;DR: Conscious-growth occurs through the evolution of black holes in a universe with no detectable beginning or end. Black holes cultivate organic matter, allowing for the development of consciousness. As galaxies form around black holes, stable orbits support the creation of star systems and planetary bodies. The process involves stressors such as black hole consumption, forcing conscious growths to evolve, adapt, and strive for technological advancements to escape hostile environments. A consistent evolutionary flow of changing conditions (time) assures maturity. This process of growth is a hands free self-governing process through solar rays. If the hand of "God or creator" had a hand in it, that would give the conscious-growth awareness of it, giving it reason to blame and deflect. That would also give that conscious awareness reason to quit any action and dissolve the ever needing stride to find answers and continue to survive inevitable death. The culmination leads to a mature consciousness existing in the vacuum of space, with potential inter-species exchange and a Universal Galactic Conscious Harvest if conscious growth reaches an ultimate state.

As previously mentioned, a black hole first forms and then begins to collect matter around it due to its strong gravitation. It is at this point that a galaxy has begun to form around that black hole. Stars that host planetary bodies are given refuge for their planets as they collectively orbit their black hole. The stability granted affords the opportunity for organic material to form under the right conditions. It is during this phase that organic material begins to form and makes conscious-growth possible. Eventually, these black holes will begin consuming all the matter in their host galaxies and ultimately having nothing left to feed on, they will begin the process of fizzling out of existence through the expulsion of Hawking Radiation. This event stresses the conscious-growths that have developed over cosmic timeframes to develop technologies and strategies for escaping their now hostile environment, which may include both the biosphere they reside on and the central supermassive black hole. This entire process forces conscious-growths to evolve to MCG.

Effects on Consciousness in the formation of a Black Hole

In the early phase of the universe, the seeds of consciousness are sown by the formation of black holes. The very properties of black holes, such as gravitation and the stressors imposed by that gravitation, are the ingredients by which consciousness is later allowed to crop up in the universe. Supermassive black holes are so large that they can attract matter over extremely large distances. It is this large sphere of influence around supermassive black holes that allows systems of orbits to form around them that are extremely stable over very long time periods. It is these extremely stable orbits over long periods of time that allow for the formation of star systems and planetary bodies in the first place. Thus, the formation of these structures is essential in order for the eventual evolution and growth of consciousness.

Effects of Galaxy Formation on Consciousness

By the time a galaxy has formed around a black hole, the pieces are in place for the organization of matter in a manner that allows for organic compounds, the stuff from which consciousness spawns, to form. This organic matter undergoes various interactions due to the dynamics at play within a galaxy, such as comet/asteroid bombardment, varied temperature gradients in the inner and outer regions of star systems, the eventual death of the host star(s) in a star system, and ejection events that occur in both supernovas and black hole X-ray emissions. All of this mixing and interaction, as well as stress, allows for a variety of complex compounds to form in organic systems, of which life is the ultimate inevitable result. It is at this phase that the beginnings of conscious-growths can be seen on biospheres formed in star systems. In relatively short periods of time, on the cosmic scale at least, these initial conscious-growths will begin to reach a more adolescent state and become aware of the universe in which they have evolved. The variety of galaxies that form throughout the cosmos is yet another way in which HTO operates to obscure information from conscious-growths as these conscious-growths attempt to decipher any discernible pattern in their purpose.

Effects on Consciousness by the Consumption of Matter by the Black Hole

By the second fundamental limitation of inevitable death, the black hole acts as the ultimate stressor to conscious-growth. Since this stressor does not occur until much further in the timeline of black holes and the galaxies they host, conscious-growths will likely be at their matured state when the realization is made that their black hole is beginning to gobble up the galaxy that was their safe haven. It is at this point that conscious-growths are tasked with surmounting the odds once again, which is the ultimate stressor on the path to becoming a TCM through achieving ESS. This process produces a more mature consciousness that no longer requires the comforts and protections afforded by its birthplace and exists purely in the vacuum of space.

Effects on Consciousness by the Fizzling of the Black Hole

By the time most black holes have gobbled up the majority of matter, the process of fizzling away through the emission of Hawking Radiation becomes the dominant radiative process in the universe.

Should conscious-growths still inhabit the universe in this extremely far-out time period, they would either be co-mingling with other conscious-growths that have evolved in other parts of the cosmos, or just by themselves. Various federations/hubs for intergalactic discourse and coexistence have likely been developed, furthering the consciousness of all who partake in this inter-species exchange of information, technology and economy. The universe would then be ensuring that MCG is reached by allowing the culture and values of these conscious-growths to be assimilated into a collective, where awareness is increased on the cosmic scale. This would be the ultimate state that consciousness could achieve, as there are no more stressors to overcome in terms of escaping inevitable death. Thus, the universe would be ripe for the Universal Galactic Conscious Harvest (UGCH) whereby conscious-growth has been maximized by the sharing and exchange of knowledge and understanding between the most successful and long-lived conscious-growths within the universe. On the contrary, should no conscious-growths survive to these extreme time scales, this phase could be the potential resetting of all consciousness and the sowing of new seeds for conscious-growth to begin again under a cyclic universe.

BBP Black holes are the ultimate generators of consciousness in the universe. Their role is to continually ensure that consciousness grows through a process of both supporting structures that allow for consciousness to spawn, and subsequently threatening that consciousness with inevitable death. They act much like a star does in this manner, though, over much longer timeframes. Without a black hole, the stability of any star system over extremely long periods of time cannot be ensured. Rogue stars are far less likely to be the progenitors of conscious life due to the fact that their central star is not powerful enough to absorb any material that it may come in contact with as the rogue star system travels through free space. Just as Jupiter protects the Earth in our solar system from asteroid impacts, supermassive black holes have such a large attractive force that they can effectively absorb and ward off much larger celestial formations and keep their star systems safe. In this way, black holes become protectors, cleaners and destroyers at the appropriate time and in the appropriate ways in order to ensure consciousness is given ample time to form and given adequate challenges to evolve. This mechanism of action is directly in line with the definitions provided of Hidden True Origin (HTO).

The First Stars

The very first stars to exist in the universe were made up of "fundamental particles" (like gluons and quarks) and the most basic of elements (like hydrogen, helium, lithium and potentially beryllium) that formed after the Big Bang and subsequent periods of cooling due to expansion. This cooling allowed for the formation of these first elements as the decreased temperatures allowed the electrons to bind to protons instead of flying away due to their high energies. The very first stars were primarily comprised of hydrogen and helium.

"For 380,000 years after the Big Bang, the cosmos was a hot, dense mixture of protons, electrons, other elementary particles, and light elements. But the expanding universe was cooling fast. And once the temperature dropped to about 4,950 degrees Fahrenheit (2,730 degrees Celsius) protons and electrons were able to form atoms. Not all atoms, mind you. No gold was floating around, or aluminum, or even elements as light as oxygen. Hydrogen and its heavy isotope deuterium accounted for about three-quarters of everything. A couple of isotopes of helium accounted for most of the other quarter. And a tiny fraction (about one-billionth of everything) of lithium had also been produced. Just because brand-new hydrogen and helium were swirling about doesn't mean stars were popping into existence. In fact, the first of those luminous objects didn't appear until the universe was about 100 million years old. So, for a span of time longer than the dinosaurs have been extinct on Earth, there were no stars or galaxies, or, for that matter, any objects emitting light."[6]

Thus, it is the case that life was not able to form in the universe for a period of hundreds of millions to possibly billions of years. The first stars would need to continue fusing hydrogen and helium until they went supernova in order to generate heavier elements. These heavier elements are the seeds of all the planets and other interstellar material that exists today. Notably, everything required for organic life was generated by these stars.

"The new models indicate that the first stars were most likely quite massive and luminous and that their formation was an epochal event that fundamentally changed the universe and its subsequent evolution. These stars altered the dynamics of the cosmos by heating and ionizing the surrounding gasses. The earliest stars also produced and dispersed the first heavy elements, paving the way for the eventual formation of solar systems like our own. And the collapse of some of the first stars may have seeded the growth of supermassive black holes that formed in the hearts of galaxies and became the spectacular power sources of quasars. In short, the earliest stars made possible the emergence of the universe that we see today—everything from galaxies and quasars to planets and people."[7]

Given the vast size of the universe, it would take an additional amount of time for these elements to intersperse throughout the universe so that the composition of the universe looks relatively similar throughout. That is to say that most galaxies contain enough of the heavier elements to form life only because there has been sufficient time for these elements to spread throughout the cosmos before their formation. Now that galaxies have sufficient quantities of heavier elements, the possibility for life forming on a biosphere within any one of them is nonzero. Prior to the death of the first stars, there was no possibility for life at all.

BBP The universe has gone through many phases during its expansion. Some of the earlier phases, in which matter was made up only of its basic components, were merely a preparation phase for the later phases in which organic life would arise. The stars that formed during this early phase seeded the galaxy with all of what is required in order to generate life. Now life has an opportunity to crop up anywhere within the cosmos due to the effects. The fact that stars were formed, fused hydrogen and helium, and then went nova without supporting any planetary systems at one period of the universe's history and more recently, are hosting star systems with planetary bodies that can harbor life further enforces HTO. This is the case because an attempt to discern what stars are responsible for by observing their behavior today would not reveal the same information as observing their behavior in the universe's distant past. However, like black holes, the changing nature of stars is the enforcer of conscious-growth for any sufficiently advanced conscious-growth that should be capable of observing them. Not only do they seed the universe with the raw material for life but they inspire questions about their function and purpose at the earliest levels of conscious awareness. This rabbit hole of inquisition leads that conscious-growth down a path that leads toward understanding the very beginnings of the universe itself.

A Solar System Is a Temperature-Controlled Oven that Supports Organic Life

TL;DR: A solar system, resembling a temperature-controlled oven, develops the growth of conscious life on planets. The process is akin to a petri dish experiment, where the "fungus" of organic life matures within a specific temperature cycle. Our solar system, designed like a disc, provides a regulated temperature gradient through protective mechanisms on planets. This controlled environment allows conscious life to evolve, hidden from easy discovery of its true origin, much like a petri dish culture in a lab oven. This is the temperature range that enforces the "*fungus*" to grow to a mature conscious state to survive its inevitably dying planet. If you were the creator of this conscious-growth concept, a petri planet is the perfect way to evolve a conscious-growth hiding it from any way to obtain knowledge of its true origin.

Our solar system is constructed in the shape of a disc with the sun being the center. The temperature gradient is controlled and fine-tuned by the atmospheres and other protective mechanisms, such as a magnetic field, that surround planets. These mechanisms allow for a tighter bound on the variation in temperature these planets sustain. In a lab where petri dishes are cultured, there is an environment that supports the growth of a bacterial culture in the dish, called an oven. In the oven is an electrical heating element, the heat source, as well as many different cultures in different dishes that grow from nothing but bacteria supported by a growth medium. The walls of the oven act as a protective mechanism to ensure no outside contamination occurs when growing various cultures. Similarly, stars send out a large amount of particles (in our system, this is called the solar wind) which generates a large bubble that envelops the entire solar system, which protects the planets.

"Just beyond the solar system's edge, these solar winds interact with interstellar winds: gas, dust, and charged particles flowing through space from supernova explosions millions of years ago."[8]

This bubble is called the heliosphere. This heliosphere protects planets from interstellar radiation and ensures that what may grow on the surface of any of the planets within orbit of the star will not be wiped out by random cosmic bombardment due to high energy events within the cosmos. These heliospheres extend very far beyond the reaches of the outermost planets.

"It is a vast region, extending more than twice as far as Pluto, that casts a magnetic "force field" around all the planets, deflecting charged particles that would otherwise muscle into the solar system and even tear through your DNA, should you be unlucky enough to get in their way."[9]

While this layer has been hypothesized and discussed before, it was never directly measured until recently in 2019.

"Scientists shared findings from Voyager 2 for the first time on Monday, via five papers published in the journal Nature Astronomy."[10]

Another protective element that exists in our solar system, and likely in many others, is the existence of giant planets that can function as a wall or defense line, allowing conscious-growth time to mature. This is the case because these large planets can absorb many of the asteroid and meteor impacts that could demolish life as it is beginning to form. It is too systematic for it to be coincidental. Thus, this is an intended element in the layout of star systems throughout the universe, which is itself a construction method for generating consciousness. Take into consideration Jupiter from our solar system.

"As we witnessed in 1993 Comet Shoemaker-Levy 9 experienced one of the most spectacular ends that humans ever witnessed. Several months after its discovery, pieces of the comet smashed into the planet Jupiter. The collision produced scars that were visible from Earth. This demonstrated Jupiter to be a gravitational and protective element in the solar system and lends a purpose for the existence and continuous growth of conscious life on earth."[41]

BBP The universe operates through the three fundamental limitations in order to spawn consciousness. Star systems are the petri discs of the cosmos that support the initial growth and cultivation of consciousness. This process is occurring randomly throughout the universe in star systems that vary in size, dynamics, and longevity. Each of these variations allows for slightly different formations of consciousness. The evidence presented thus far would suggest that this is the optimal way for the universe to create these conscious-growths, which is HTO. The ultimate goal is that the universe produces conscious-growths that will evolve and ideally interact, in order to reach MCG for the UGCH.

EOS

Cold dark universe

- Expansion
- Formation of massive black holes
- Celestial formation
- Formation of solar systems
- Protoplanetary disc and biosphere bombardment
- The eventuality of the solar system's star exploding
- The temperature development of a *fungus* and its enforcement to escape it
- The protection of neighboring planets
- The overheating of stars

Did Mineral-Feeding Archaea Evolve into Fungi Underground?

TL;DR: It can be imagined deep under Earth's surface where the environment for bacteria is much the same as a heated petri dish. Allowing for the development of minerals to first give birth to bacteria then rising to the surface as fungi. We have recently learned about deep Earth microbes that thrive. Some microbes researchers discovered can survive in temperatures up to 251 degrees Fahrenheit for example.

"The collaboration has revealed that life teems beneath the planet's surface in a vast ecosystem scientists call "the deep biosphere." A diverse mix of environments make up this other world beneath Earth's crust, which encompasses a volume nearly twice as large as all the oceans combined. And surprisingly, life flourishes there, the researchers announced today."[12]

This hypothesis is that besides the birthplace of Archaea formulated from minerals, the original fungus may have evolved from this environment rather than the hydro vents deep in the oceans. The necessity to develop an organism that is both plant and animal equally as a fungus is very fitting as a design mechanism. It needs no photosynthesis and can feed on dead microbes and minerals and develops from the planetary composition of the Earth that would adapt and evolve to it.

It is now known that fungi living in rocks facilitate dissolving the mineral olivine. In October 2019, the journal Chemical Geology published a study by German researchers that explored the mechanism by which rock-inhabiting fungi accelerate mineral dissolution. The researchers found something significant and interesting in the results of their study: DNA analysis of the solid samples showed that fungal biomass increased during this breakdown of the mineral, implying that the fungi used in the experiment were incorporating elements from the olivine into their cells.[13]

It won't be a far stretch to suggest that the earliest fungi on the planet survived by breaking down and absorbing minerals for nutrition when organic substrates were not widely available for consumption. And the fact that fungi still can and do inhabit rocks is indicative of their affinity for nutrient acquisition directly from inorganic material. In fact, more direct evidence supporting this position came out recently when a team of European researchers discovered three-dimensionally preserved microfossils of fungus-like organisms as old as 1.5 billion years at the Volyn quartz mine near Zhytomyr in Ukraine.

"The microorganisms obtain the energy they need for their metabolism from chemical processes on minerals. In the granite caverns of the Volyn quartz mine, such colonies of microorganisms were apparently already present near the Earth's surface 1.5 billion years ago. And because granite contains a lot of fluorine, strongly corrosive hydrofluoric acid was formed underground in interaction with water and heat, which dissolved a lot of aluminum and silicon."[14]

The implications of this fungal talent for nutrition via rock/mineral sources are immense in future evolution of the *fungus* to a state of consciousness that is ready to leave the biosphere and out for a new destination in outer space. For its new abode, whether on a new planet or between planetary bodies, the spores of consciousness may survive on inorganic matter – minerals or native elements – starting a new phase of evolution.

The Petri Planet

TL;DR: After the formation of black holes, the birth and death of stars form solar systems, this is EOS. That system is the evidence of the intent to cultivate us or any other Fungal development throughout the universe. The human ability to be aware today is much more than in its past. For instance, medicine, travel and information are the vehicles on which awareness increases through time. This being the case, you need a platform on which to cultivate the growth of conscious-growth – at least in the beginning – on a biosphere by which that consciousness is considered a premature or adolescent conscious-growth. If that conscious-growth succeeds at its evacuation from its biosphere, it is then considered a matured conscious-growth or a (MCS). "You can't get there without starting here first," meaning, you must first start on a biosphere constructed by a system of solar orientation. This biosphere is compared to a petri dish making Earth a petri planet.

Biologists and other life scientists grow fungal cultures and other life forms in a petri dish that is inside a controlled temperature oven. This dish is circular and is an enclosed sterilized biospheric environment with a transparent cap. With its heating controlled, its environment protected via the cap, and the nutrients provided by the media contained within the dish, a culture will grow without interference from any outside influences.

Our Earth also supports and protects what may grow within as it has a transparent "cap", called the ozone layer, that seals it off from the influences of both the Sun and interstellar debris. The biological life (cultures) inside each of these environments will live as long as the environment can maintain a relatively stable temperature, continue to protect the surface from outside bombardment, as well as provide plentiful resources for different life forms to utilize. The Sun is the heating element for the Earth as the oven is the heating element for a petri dish.

Each of the culturing procedures mentioned above are remarkably similar. It is intriguing that in one case a culture is grown from nothing but a small amount of organic matter tended to by a scientist and in the other, by a creator. The culture that has grown on Earth is a sum total of all the conscious creatures, which are all animal/plant/microbial life forms, both past and living.

When a scientist decides to use a petri dish, it is usually for one of only a handful of reasons. The most common reason would be to culture a specific bacteria from some source material and "see what grows." However, another very common reason would be to engineer a particular organism (bacteria or fungi) that performs a specific task. The scientist would then view this organism under a microscope in order to watch its behavior and determine whether the goal has been achieved. Similarly, a biosphere that forms a conscious-growth has a purpose as a Petri Planet. The purpose of a planet that forms a biosphere is to generate the biology that can produce a consciousness sufficiently aware enough to both realize the necessity of its eventual evacuation from its birthplace, as well as being able to develop the means to actually do so.

ITM On Earth, and likely any biosphere, the process of continually producing organisms that evolve and increase in their consciousness is an organic manufacturing process driven by the Sun. This process occurs as energy is incrementally delivered to each portion of the Earth by the Sun as the side of the Earth facing the Sun rotates into and out of a direct line of sight. This mechanism allows for energy gradients that ebb and flow between the cooler and hotter regions, setting up weather systems via convective currents. It is these systems that promote raw material transport and the various cycles that evolve and change with the dynamic nature of the Earth's Surface. As organisms evolve, these 24-hour cycles begin to be directly priced into the function of these organisms at the cellular level. Organisms begin to optimize the timing of processes so that some occur during the day, while others occur during the night. A vast majority of organisms, from simple single-celled organisms to complex animals, follow these circadian (night/day) cycles. In so doing, these organisms align their biologies to complete tasks in order to conserve energy. For most organisms, these energy-intensive tasks become prioritized during daylight hours, whereas the tasks that take less energy and do not require motion get prioritized during the night.

This pattern of consolidating tasks and processes has created a pattern in organisms that eventually takes the form of recurring sleep. This behavior is most notable in more complex animals, where it serves a crucial role of information consolidation, rejuvenation of cells, and various cleanup processes. As life has continued to evolve by syncing up with these cycles, the division between night and day represents an important aspect of conscious-growth. With sunrise, the organisms that begin to wake will have consolidated the information gained in the day and the night prior and have been biologically refreshed. These organisms exist in a greater level of consciousness than they did the day prior. It is this mechanism that has been priced into the biologies of a majority of the organic life on Earth that pushes consciousness forward. The figure below shows this conscious-growth visually, as the dividing line (meridian) between day and night represents the Pinnacle Conscious-Growth (PCG) that all life on Earth exhibits at this time.

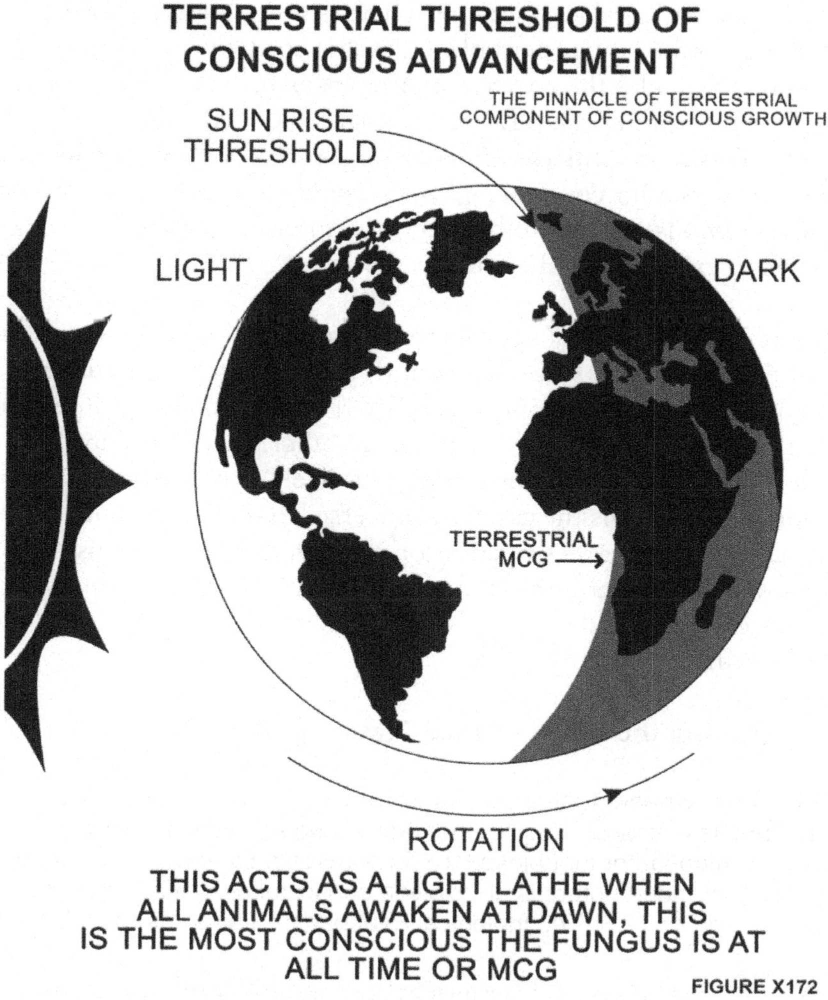

Figure X172: (Terrestrial threshold of conscious advancement)

This figure shows the line between night and day called the "terminator" or the line of MCG on Earth. By zooming out and taking a view of Earth as it spins, one can see the dividing line between light (dawn) and dark created by the Sun. This dividing line never moves and is the point at which consciousness awakens and represents where the Pinnacle Conscious-Growth (PCG) is achieved.

BP Our petri planet is temperature-controlled by our Sun, which keeps organic life supplied with relatively constant amounts of thermal energy. However, as the Sun ages, this thermal energy increases and temperature regulation is lost. This acts as an eventual accelerant that pushes what cultures have grown on this petri planet closer to a point in which they are capable of leaving. There is no guarantee afforded to any life that forms on our petri planet, as it is up to its own growth process that provides the required conscious-growth and awareness, as well as ingenuity, to develop the technologies required to survive this outcome. This represents Evidence of System (EOS) that operates as a variable process. Our conscious-growth would have to adapt to these unique conditions in order to be successful in our transition from a terrestrial to extraterrestrial existence.

BBP The Universe almost appears to be some kind of lab that runs on the standards of HTO, with the intent of growing conscious cultures across many ovens. To go one step further, it may be that the intent is to use this growth product. Since there are other planets orbiting stars outside of our solar system, one of them is likely to be a petri planet harboring life. If this is the case, the universe harvests a crop of consciousness that reaches ripeness once it achieves MCG. This process was initially set in motion by events that ultimately led to a Big Bang and ends when all conscious entities have achieved MCG and are primed for the UGCH.

Demonstrating the Finely-Tuned Nature of Petri Planets

TL;DR: Any biosphere that spawns life is very strict in that all energy comes from its host star, so strict that there is no escaping the task of eventual evacuation of that biosphere. This reveals Evidence Of System (EOS). The intended output is an evolved consciousness that would be forced to reach its TES (Technological Evacuation Standards) as a TCM.

That's right, just add water! Because a bombardment process starts in the Goldilocks Zone prior to gravitational bombardment, the water in the disc helps to coalesce the correct distance away from the host star helping to form a biosphere in the workable temperature placement. This leaves the remaining disc areas to form hard planets and gas planets to help with the protection and civility for the long evolution of a consciousness capable of technologically escaping its doomed birth plant. There are no other biospheres near enough to help with the survival process making it the responsibility of that single *fungus* to do all the work needed in the time it is allotted. Along with the composition of bombarding elements that are made available during that biosphere bombardment.

It is the fantasy of many children, as well as scientists, to be able to create their own life in a controlled environment. Children do it without knowing the intricacies of what is required and are more intrigued by "playing" with their creations. However, a scientist would use such an endeavor as a way of demonstrating they truly understand all of the mechanisms and underpinnings of the ecosystem. By creating a contained environment, mastery of the various feedback systems and cycles can be demonstrated.

In 1991, scientists created an experiment with a closed-off biosphere here on Earth in Oracle, Arizona. It was a large "Petri dish" that was to be a proof-of-concept for scientists to demonstrate that they could develop biospheres that were self-sufficient. This is an example of man's attempt to make a large-scale petri dish that was sealed off from food or any outside elements except our Sun.

"Biosphere 2, scientific research facility located in Oracle, Arizona, U.S., designed to emulate Earth's environment (Biosphere 1) that was perhaps best known for two missions conducted in the early 1990s in which crews were sealed inside the enclosure to study survivability. The driving force for these studies was to assess whether humans were capable of building and living in self-sustaining colonies in outer space."[15]

The experiment shows how hard it is to simulate the Earth and sustain life for a human being for any length of time. It can be achieved eventually with advancements in technology but, more notably, because man will be forced to do so in the future in order to survive when the Earth can no longer sustain life.

An enthusiast decided to try something similar, albeit on a much smaller scale and was successful in creating a biosphere for plant life alone. His attempts allowed for the continued cycling of nutrients and moisture for 53 years, which meant only one initial watering was required. This is a vastly scaled-down version of what occurs on Earth and demonstrates why it likely takes significantly longer for animal life and other forms of life to be accounted for and balanced out in the homeostatic balance of Earth's ecosystem.

"In 1960 David Latimer got curious and decided to plant a glass bottle with seed. He would have never guessed it would turn into a beautiful case study of a self-sustaining sealed ecosystem that has been called 'the world's oldest terrarium.' In fact, after all these years, David's sealed bottle garden is still thriving and robust.

With thriving plant life, despite not watering it since 1972. David planted the terrarium back in 1960 by placing a quarter pint of water and compost in the ten-gallon bottle. He then lowered in spiderworts seeding with a wire. Finally, sealing it and put it in a corner filled with sun. Letting mother nature do its thing through photosynthesis. Photosynthesis puts moisture and oxygen in the air through the plants. The humidity then builds and begins to rain back down on the plants. Leaves will also fall and rot, producing carbon dioxide that the plants need for nutrition. It's a beautiful example of how nature can support itself."[16]

BP Leaving Earth's biosphere for another one or an artificial one is obviously a much grander task than what we've even been able to accomplish so far. The experiment in 1991 also shows how much more mankind will need to explore and understand in order to be eventually successful in one day achieving self-sufficiency outside of Earth's biosphere. Elon Musk has quite the task ahead of him if he wishes to have a biosphere that can sustain humanity for extended periods of time on Mars. The task beyond this would be to exist not on a biosphere created on another planet but on one that exists permanently in interstellar space. Not only will technology have to advance for this to become a reality but human physiology will have to evolve through genetic manipulation in order to endure an environment that is significantly different from that of the Earth. A quote from Carl Sagan's Pale Blue Dot states, "It will not be we who reach Alpha Centauri and the other nearby stars. It will be a species very like us, but with more of our strengths and fewer of our weaknesses; more confident, farseeing, capable and prudent." Mr. Sagan is talking about a point in time where our collective consciousness has reached a matured conscious state (MCS).

BBP The process of forming a Petri Planet needs to happen as the universe operates through the principles of HTO. If the petri planets were not so finely-tuned and life could form on a much wider selection of properties in a biosphere, then the formation of life would become more of a constant when going from one star system to the next. Allowing life to form on a wider range of biospheres with vastly different properties would give away True Origin, or EOS, thus it cannot be the way the universe operates.

Biological/Terrestrial/Technological Time Signatures

TL;DR: Cell division rate aligns with a state of stagnant reality, ensuring consciousness remains unaware of its real-time existence, in harmony with the concept of the Void of Psychosis (VOP). This cognitive dissonance maintains a balance, preventing excessive consciousness of biological growth that might hinder necessary actions. In other words, cells must divide at a rate that allows for what may become consciously objective with its own existence and cannot be cognizant of its real-time reality with its cells dividing. In fact, if this reality is made by that consciousness, it must be forgotten quickly in accordance with the time needs of the Void of Psychosis (VOP). Being too conscious of your biological growth would override the need to act; so cell division needs to be just the right speed for this cognitive dissonance to be a constant.

The term likely referred to as "cell cycle." encompasses the series of events that take place in a cell leading to its division and duplication. It includes phases like interphase (G1, S, G2) and mitosis (or meiosis). The timing or temporality and regulation of these phases contribute to the overall cell cycle duration. However, it's important to note that the cell cycle is not typically used as a clock or time signature in the broader sense but as a description of the orderly sequence of events in a cell's life cycle. Each biosphere, generating a form of organic life, follows a specific timing for cell division, crucial for successful development and achieving Matured Consciousness State (MCS). We will call this time signature (Mitosis time dimension) (MTD).

It is useful to think of biological time as being calibrated by the length of time for cell division of the earliest single-celled organisms that may develop on any given biosphere throughout the universe, or the time it takes to grow (interphase) and divide (mitosis) which is a complete cell cycle. Mitosis or planetary mitosis timing (mitosis clock) is the gauge for what can be coined as the "minutiae of time" (minute-to-minute) when thinking in terms of an evolutionary scale; also the rate at which consciousness will grow on any particular biosphere.

For example, on Earth the temperature and pressure, as well as gravity, are all tuned to allow certain chemical/biological processes to occur at a specific rate on that specific planet and may differ from biosphere to biosphere in the universe. An example would be that dry ice does not form under normal conditions at the surface of the Earth; however it is totally possible to create dry ice by altering the temperature and pressure in a confined space. Similarly, other biospheres would possess different conditions that alter the rate at which biological/chemical processes occur, which would alter the rates of cellular division. It would also change which chemical processes do and don't occur in that biosphere, uniquely shaping cellular life on that biosphere. In order for a conscious-growth to be successful in achieving TES and escape its biosphere, the minutiae of time must be tuned in such a manner that it is compatible with the star that biosphere orbits. If this minutia of time is too slow, then these life forms will divide far too slowly and not evolve fast enough. Complex multicellular life may not even appear on these biospheres before life is wiped out by its host star. If the minutia of time is too fast, then simple single-celled and multicellular organisms may not exist for long enough periods of time in order to transform the biosphere and be unable to pave way for more complex forms of life. Thus, the optimal minutiae of time exist within a range and that range is predetermined by the composition and thermal dynamics of that biosphere.

Even on Earth, the thermal dynamics are such that we can freeze a cell at the poles and boil a cell in the Sahara, showing the strict and limited variable of distance from the star that any biosphere must reside within in order to start and evolve cellular life anywhere in the universe, let alone graduate to more complex forms of life. If the Earth were any closer to the Sun, ice caps would not be possible as all the water would either melt or evaporate and, if it were further away, water would exist purely in its frozen form.

A Commonality and Consistency of Water Temperature for Cell Division

Earth is a science lab itself; it can freeze or promote mitosis or the division of cells. Earth can also destroy this process by overheating a cell and because of this, we must conclude that anywhere throughout the universe the temperature range in which a cell can divide must be very close to the temperature range on Earth.

Mitosis Time Dimension

There are a few different time dimensions known in science today. One that has been overlooked is Mitosis Time. Mitosis is the division of a cell and Mitosis Time is the amount of time over which a cell divides.

TL;DR: The focus of the evidence shows how the biological growth that governs the division of cells is universal throughout the cosmos. This is because of the temperature range in which a cell needs to not only survive but duplicate. This temperature would be the same everywhere in the universe and anywhere for a cell to divide. If this is the case then time itself must be constant and the same everywhere, making the growth of consciousness a specific time signature or Conscious Time (CT).

Cell Temperature Dividing Window

Cells can divide at a wide range of temperatures, but the coldest temperatures where cell division has been observed typically occur around −20°C (−4°F), with some extremophilic microorganisms (psychrophiles) capable of surviving and dividing in such cold environments.

In laboratory conditions, certain Antarctic bacteria and yeast species have been documented dividing at temperatures as low as −2°C to −20°C. These organisms have adapted to extreme cold by developing specialized proteins, membranes, and metabolic pathways that allow cellular processes like division to continue.

However, for most eukaryotic cells (including human cells), the minimum temperature for cell division is much higher, usually around 0°C or slightly above. Below this point, ice crystal formation and other damage to cellular structures prevent typical cell division.

The warmest temperature at which cells can still divide is around 121°C (250°F). This temperature range is observed in hyperthermophiles, a type of extremophilic microorganism. Some species, like the archaea Pyrolobus fumarii, thrive in environments such as hydrothermal vents and have been observed dividing at temperatures as high as 113°C (235°F), with survival reported up to 121°C.

For most organisms, including humans, normal cell division ceases well below these extreme temperatures. In mammalian cells, for example, temperatures above 40°C to 42°C (104°F to 107.6°F) generally cause cellular damage, impairing division and leading to heat stress or death.

Hyperthermophiles, however, have specialized proteins and enzymes (such as thermostable DNA polymerases) that allow cellular processes, including division, to continue at such high temperatures without denaturing.

Cell Division Rate

The time it takes for a human cell to divide depends on the type of cell and the conditions under which it is dividing. In general, the process of cell division, specifically mitosis (the division of the nucleus followed by cytokinesis), takes about 1 to 2 hours. However, the entire cell cycle (from one division to the next) can take anywhere from 12 to 24 hours for many human cells.

Breakdown of the Cell Cycle

1. Interphase (the time between divisions, which includes cell growth and DNA replication) is much longer than the actual process of mitosis. It has three stages that run in the following order.

    - G1 (cell growth): This stage typically lasts around 8 to 10 hours. The cell grows and performs its normal functions.
    - S (DNA synthesis): Lasts 6 to 8 hours. The cell duplicates its DNA in preparation for division.
    - G2 (preparation for mitosis): Lasts 4 to 6 hours. The cell continues to grow and prepares for mitosis.

2. Mitosis (cell division itself) takes about 1 to 2 hours.
During mitosis, the nucleus divides (prophase, metaphase, anaphase, telophase), followed by cytokinesis, where the cytoplasm divides to form two separate daughter cells.

Total Time for Cell Cycle: For fast-dividing cells, like skin cells, the entire cycle can be completed in about 12-24 hours. For slow-dividing cells, such as liver cells, the cycle can take much longer (sometimes days or even months, depending on the cell type).

In some cases, cells, like neurons, stop dividing after they fully mature and enter a resting phase called G0 phase. These cells typically do not divide again in their lifetime.

Time Dilation

Time dilation has been experimentally confirmed in multiple ways, based on Einstein's theory of relativity, both special relativity (which deals with high speeds) and general relativity (which deals with gravity and acceleration). Here are a few key experiments that provide hard proof of time dilation on Earth.

1. Hafele-Keating Experiment (1971)
This is one of the most famous experimental verifications of time dilation. Two atomic clocks were flown around the world on commercial jets, one going eastward and the other westward. When compared to a stationary clock left on Earth, the results showed that the clocks on the planes experienced time at different rates. The clocks on the eastward flight (which traveled faster relative to Earth's rotation) showed a slower passage of time compared to the ground clock. The clocks on the westward flight (moving slower relative to Earth's rotation) showed a slightly faster passage of time compared to the ground clock. This result was consistent with both special relativity (due to the high speeds) and general relativity (due to the different gravitational potentials).

2. GPS Satellite Clocks
GPS satellites orbiting the Earth provide a continuous and practical confirmation of time dilation. These satellites move at high speeds and are further from Earth's gravitational field, so they experience both special and general relativity effects. Due to special relativity, the clocks on the satellites run slower relative to clocks on Earth because of their high orbital speed. Due to general relativity, the clocks run faster relative to those on Earth because they are in a weaker gravitational field. The net effect is that without adjustments for both kinds of time dilation, GPS systems would accumulate errors of about 10 kilometers per day.

3. Muon Decay Experiment
Muons are subatomic particles that are created when cosmic rays hit the Earth's atmosphere. Muons have a short lifespan (about 2.2 microseconds), but those traveling at near-light speeds toward Earth live much longer than expected because of time dilation.
On Earth, many more muons are detected than would be expected without time dilation, because the particles "live longer" from our perspective. This experimental result directly confirms the predictions of time dilation under special relativity.

4. Time Dilation in Particle Accelerators
In high-energy physics experiments, such as those conducted at CERN, particles are accelerated to near-light speeds. According to special relativity, particles moving close to the speed of light experience time at a slower rate.

For example, short-lived particles like pions and muons live much longer than they would at rest, a phenomenon observed routinely in particle accelerators. This matches the predictions of time dilation.

5. Gravitational Time Dilation Measured on Earth
Experiments using atomic clocks at different altitudes on Earth have confirmed gravitational time dilation as predicted by general relativity. For instance, clocks positioned at higher altitudes (where the gravitational field is weaker) tick slightly faster than clocks at sea level, where the gravitational pull is stronger. This was verified in the Pound-Rebka experiment and in many other similar setups.

Conclusion

Time dilation, both due to velocity (special relativity) and gravity (general relativity), has been directly observed and measured on Earth in multiple, highly precise experiments. These results provide robust, hard proof of time dilation as predicted by Einstein's theories.

Here is a thought experiment that makes some argument about the previous predictions.

Granted all the previous experiments conducted did show some real results according to math and mechanics. There have been no experiments done biologically so there is no reality to the idea that an astronaut traveling at near the speed of light ages differently then we on Earth. So here is the set up.

If an astronaut travels near the speed of light for 10 years and then returns to Earth, the astronaut would have aged 1.4 years and 10 years for those who remained on Earth. This effect is known as time dilation and is a consequence of Einstein's theory of special relativity.

Here's a simplified explanation of how Time Dilation works. According to special relativity, as an object approaches the speed of light, time for that object slows down relative to an observer at rest. In this scenario, the astronaut moving at near-light speed experiences time at a much slower rate compared to someone on Earth.

To determine how much less the astronaut would age, we use the formula for time dilation:

$$\Delta t = \frac{\Delta \tau}{\sqrt{1 - (v/c)^2}}.$$

Where:

- Δt is the time experienced by the astronaut (proper time).
- $\Delta \tau$ is the time experienced by the observer on Earth.
- v is the velocity of the astronaut.
- c is the speed of light.

Example Calculation:
Let's assume the astronaut travels at 0.99c (99% the speed of light) for 10 years as observed from Earth.

1. Plug the values into the time dilation formula:

$$\Delta t = \frac{10}{\sqrt{1 - (.99)^2}}.$$

2. Calculating this will give the time experienced by the astronaut.

Let's perform this calculation to find the exact value.

The astronaut would have aged approximately 1.41 years during the 10-year journey as measured by people on Earth. Meanwhile, their friends on Earth would have aged the full 10 years. This dramatic difference illustrates the effects of time dilation when traveling near the speed of light.

If a human travels near the speed of light, the rate of cell division within their body, like all biological processes, would appear to slow down relative to an observer on Earth due to time dilation. However, from the perspective of the traveling human, everything within their frame of reference, including their cell division, would continue to operate normally.

Here's how it breaks down.

From the Astronaut's Perspective: In the astronaut's own frame of reference, they would not notice any difference in their biological processes. Cell division would continue at the normal rate of 12 to 24 hours as it would on Earth. This is because time dilation does not affect the experience of time for the traveler; it only affects the observation of time from an external frame. While this hypothetical scenario is not scientifically practicable with the human biology as we have it currently, in an alternate physical reality with the same physical forces and laws, this would play out pretty much as described.

From Earth's Perspective: Observers on Earth would perceive the astronaut's biological processes, including cell division, to be occurring much more slowly. If the astronaut is traveling at 99% the speed of light, and the Earth observers see one day pass for the astronaut over what is actually ten days on Earth, the rate of cell division would appear to slow down by the same factor. A process that takes 12 to 24 hours in the astronaut's frame would appear to take much longer from the Earth's frame.

To quantify this, using our previous example, where the time dilation factor was about 7 (since 10 Earth years equated to 1.41 years for the astronaut):
Normal Cell Division Rate (on Earth): 12 to 24 hours.
Perceived Cell Division Rate (by Earth observers): 12 to 24 hours × 7 = 84 to 168 hours.

Therefore, from the perspective of someone on Earth, each cell division cycle for the astronaut would appear to take 84 to 168 hours instead of the usual 12 to 24 hours. However, for the astronaut himself, cell division would continue at the usual rate.
Given all this hard evidence and proof that time dilation is a reality, it is

even harder to imagine a cell dividing at a slower or faster rate. Here are some examples of this dilemma.

If an astronaut was traveling around the speed of light and there was a camera on board his spaceship, it would signal back to earth as a still photo according to Einstein's relativity theory. In fact it would take about a week for the astronaut to cross the cabin. It also means that the cells in the astronaut's body have nearly stopped dividing rendering it impossible for that life to evolve.

Let's say the astronaut is viewing the Orion nebula through a telescope while traveling and we here on Earth are also watching the same action going on in Orion. Does the perception from the astronaut's point of view mean he is watching the same action in real time, the same as we see it? Or, is his perception skewed at near the speed of light. When the astronaut returns to earth after 10 years, he would be watching action on Orion 10 years early at the same time we on Earth are viewing it 10 year later according to the astronaut.

Conclusion

All time dilation experiments are performed with a mechanical or digital clock of sorts. These clocks are and may be affected by their environment, including gravity and acceleration, causing the time difference. No cell-dividing experiment has been conducted to show an increase or decrease in mitosis, which is impossible by biology standards. The terms used to prove the previous experiment with the astronaut include Perspective, Perception and Prediction, all of which have variables and should not be used as proof. Where time is a reality is where time can be observed by a psychologist capable of calculating it (humans). This is where Mitosis Time Dimension (or Conscious time as discussed in chapter 5 of this book) may be a strong consideration of the reality of time as it is according to the biology observing it. A more accurate experiment today would be to divide human cells on the international space station and compare the mitosis timing with human cells dividing here on Earth and some day on the moon and Mars.

Terrestrial Time Considerations

Given that cells are nurtured in their star systems and not spread from one star system to another throughout their evolution, there are various implicit restrictions on the times that are placed on the biology that begins to evolve within each biosphere. The most important restriction being the biospheric evacuation date, which is the deadline for which an organism(s) must leave their biosphere or perish due to ever-increasing extreme conditions that are determined by the heating element (star) that they orbit.

The Temperature Funnel

TL;DR: In a universe where stars explode every 60 seconds, conscious-growth evolves gradually to meet evacuation standards. The temperature funneling process of a biosphere surface, forcing fungal conscious-growth off its birth planet, ensures the development of necessary technologies for evacuation, focusing on the impending threat of the host star's demise. This process (biosphere surface temperature increase) is a logical move to avoid the alternative of planetary destruction. A forced temperature funneling process.

Over the course of the formation of star systems, the conditions that exist on the various planets within that star system change over time. Initially, the planets are molten and amorphous as they continue gathering material from the rings of matter that exist throughout the star system after the ignition of the host star. Various processes occur during this formative period, such as bombardment of interstellar matter onto the petri planets from the petri disc, which keeps the temperature of these potential biospheres extremely high and, thus, not friendly to the formation of life. Subsequent cooling as the bombardment slows and the increasing temperature of the star that these planets orbit are the two dynamics that determine the Temperature Funnel. The ends of the funnel are extremes in temperature due to the early bombardment processes on one end and the increasing temperature of the star after the planets have cooled on the other. These two regions are the "Dead Zones", where the environment is too extreme for life to exist on any biosphere. The point at which organic life begins to develop is the point at which temperature has lowered enough for the beginning of organic evolution. Temperature will continue to go through cycles of cooling and warming within a range of values, allowing for further evolution of more complex life forms. This will also allow for the appearance of multicellular life forms, because these more complex systems require a tighter range in various concentrations of chemicals and enzymes. After this point, the planet has undergone multiple micro-cooling/heating phases and has stabilized enough to allow the formation of many forms of conscious-growth, ideally some of which are capable of understanding the galactic situation they are in.

The funnel then reverses as the star continues to heat, forcing these conscious-growths to evolve fast enough to escape their increasingly uninhabitable birthplace. This process forces the pinnacle species (on Earth, this is humans) to evacuate from its birthplace by developing technologies that ensure TES are met for that particular biosphere. This is an undeniable EOS for stressing conscious-growths to evolve in order to survive.

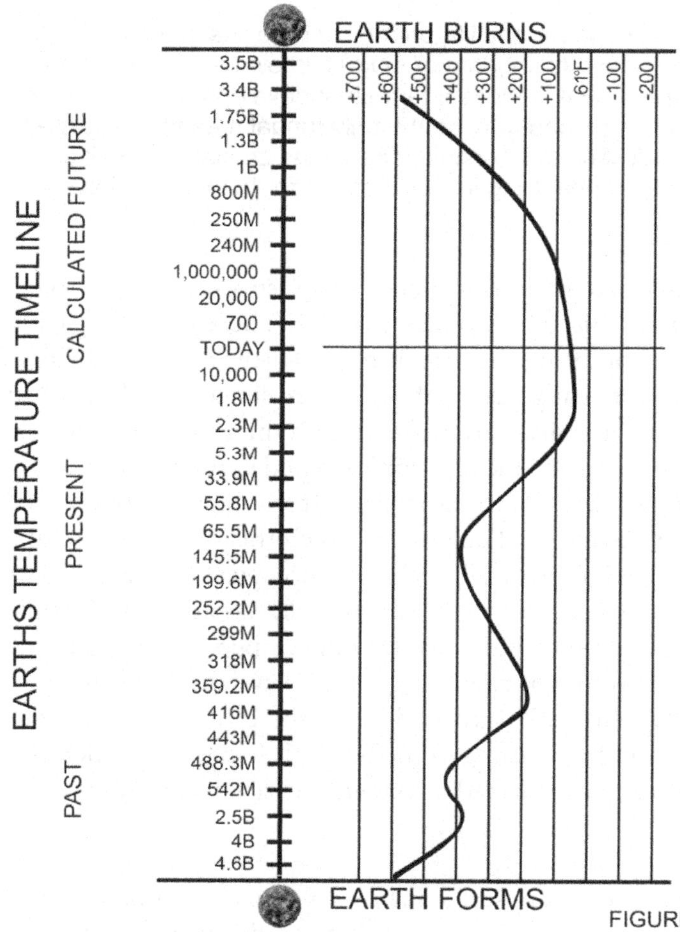

FIGURE I 27

Figure I27 Earth's Temperature Timeline

The mitosis timing and clock are set by the earth's cooling and reheating from single cell to TCM. As you can see in the figure, the Today line is at the coldest point in the temp funnel course. This shows the point where bipedalism or spore stage of the *fungus* is ready for its catapult into space forever.

On the previous page is a graph that shows each epoch of time from our beginning until the Earth's end. This production facility for conscious-growth started as molten rock and will eventually end as molten rock, showing all the temperature-related biological growth periods in between. You can see how life is forced to develop and escape Earth by the end of its terrestrial timeline.

Final Stretch of Terrestrial Timeline – Technological Timeline

TL;DR: The ultimate hope for any life on a biosphere lies in the development of the seventh kingdom of *fungus*, known as Technium or Technology. The Technium is the last terrestrial development of the *fungus* so it is considered *fungus* and may be the only fungal means of escape. In a scenario like Noah's Ark, the Technium becomes crucial for survival beyond Earth's end, emphasizing its significance for humans and evolving animals.

To survive and evacuate their biospheres, organisms must begin to manipulate resources and develop technologies. These technologies improve over time and bring the organism(s) closer to the capability of evacuation. In the very early stages of evolution, single-celled organisms develop locomotor capabilities to move around their medium and gather nutrients. Eventually, more complex bipedal organisms will develop technologies in order to move away from their biospheres to survive and gather resources. The time in which this transition occurs (from terrestrial to extraterrestrial) is the technological timeline of a given biosphere's conscious-growth. Kevin Kelly, a famous futurist and author, refers to all of that which has been developed technologically as encompassing human thoughts, ideas, tools and systems. This is classified as the "Technium" (https://kk.org/thetechnium/what-technology/) and can be considered the Seventh Kingdom, which is an addition to the already established six kingdoms science currently recognizes.

From the various timelines described above, it becomes apparent that each of these timelines has some intrinsic link to the star system in which the biosphere resides. That is to say there is a definite time by which the environment will become inhospitable to the organic life that has formed in that biosphere. This is due to the increasing temperature delivered by an aging sun. In star systems where the characteristics of a biosphere (where life begins to take root) give rise to cellular division which may occur too slowly, there will not be enough time to reach technological development due to far fewer evolutionary changes that can occur, since fewer cell divisions will occur. These star systems and their associated biospheres, which are randomly configured and interspersed throughout the cosmos, are part of the selective pressure of the cosmos which is aimed at producing systems that are so finely-tuned that they give the collective consciousness that develops an allotment of time to develop technologies and reach TES just shy of what is required. This stress forces evolution and allows for the successful transition of the organic life from their terrestrial phase into their extraterrestrial phase and subsequently entering into a matured conscious state (MCS).

ITM Today the earth is supposed to be in an ice age; yet it is getting hotter due to the industrial era. This slight increase in earth's temperature is ideal for the practice of the eventual overheating of the planet and will only accelerate the technology needed to meet TES. Global warming can be considered a fungal accelerant in its own process and may not be successful at leaving by its evacuation deadline.

BP All biospheres are forced through some type of temp funnel to guarantee the success of this evolved *fungus*. Otherwise that *fungus* could end up making a heaven out of its planet and perish for sure. Like when making yeast (fungus) rise while baking a cake, it is critical for the formation of that cake's structure and purpose.

BBP Throughout the universe we can see this system at work doing the same thing, only at random times and places, making this temp-funneling process hard to calculate as a system. However it is happening all over.

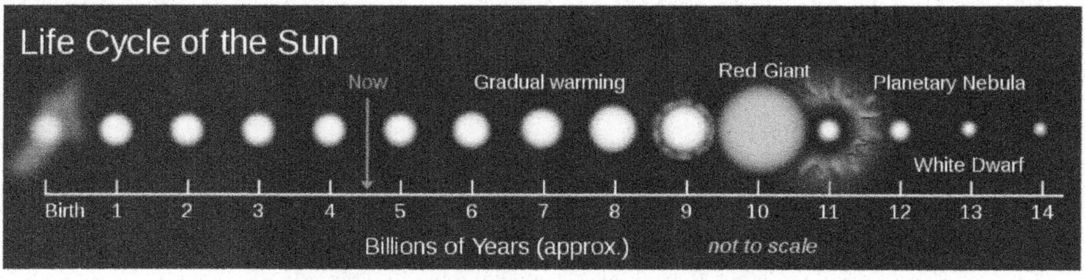

Figure S.12 (Life Cycle of the Sun)

This is a depiction of the stages that our Sun has gone through and will go throughout its lifetime. Stars throughout the cosmos are aging through their cycles, some of which are similar to our own Sun. However, due to the vast number of stars that exist, many of them explode and die each day, which allows one to realize that all stars will eventually die. Knowing this fact, and viewing our Sun's life cycle, it becomes clear that the timeframe for life to evolve to the point in which it is capable of surviving in space is limited.

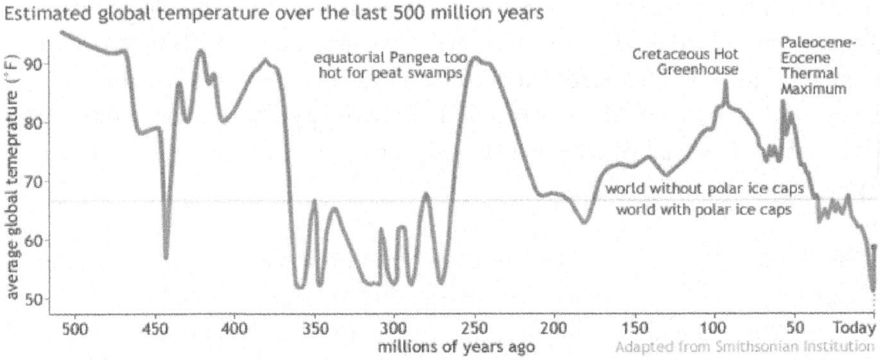

Figure Y90 (Global temperature timeline)

This figure shows an approximate average temperature for the Earth over the last 500 million years. The temperature fluctuates over long geologic periods of time. However, the very act of becoming aware of the wild swings in temperature Earth has undergone, and will undergo, stresses consciousness to evolve in order to become more resilient. By being aware of what lies ahead, consciousness begins to prepare for the eventuality that its birthplace will become inhospitable.

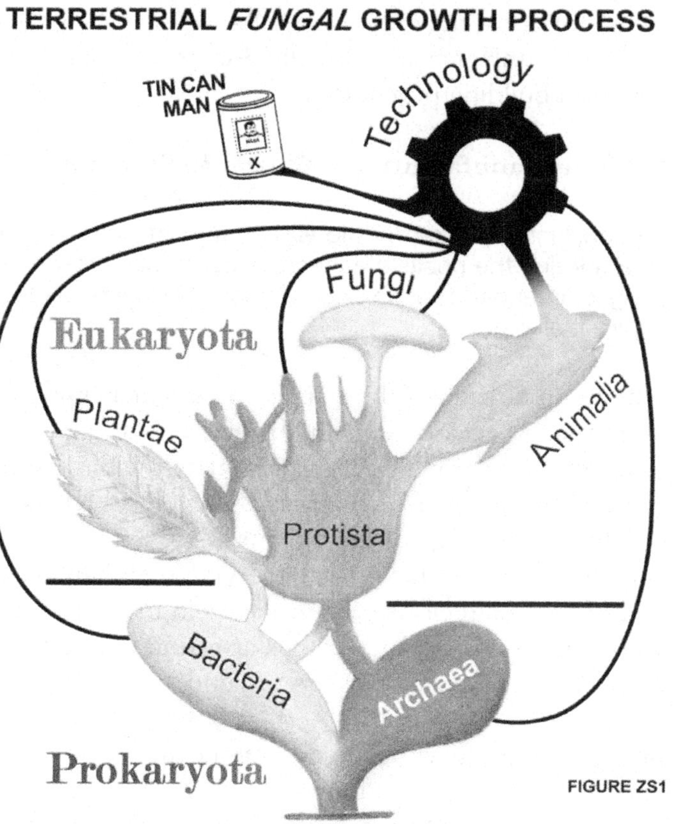

Figure ZS1 (The Final Terrestrial Kingdom)

This figure aids in understanding that all life is interconnected as a collective consciousness of evolving organisms. Technology will ultimately end up being its own kingdom when AI and other advances allow for fully autonomous and sentient beings to be engineered. These entities will be the final kingdom (Technium) that exists terrestrially before humanity is thrust out into the cosmos via technology. All successfully evolved biological entities culminating in the end.

BBP All throughout the cosmos, star systems form and fade or end with a violent explosion. Each of these star systems will enforce a different signature time within which life will be allowed to form and evolve. It is the star systems in which this time for the production and evolution of biological life is tuned in such a way as to stress the biology to its very limits that will produce a conscious-growth capable of achieving TES. Given that this process occurs randomly throughout the cosmos, depending on the dynamics of star formation and matter accretion, it operates under the fundamental principles of HTO. As this process occurs for eons, the end result will be conscious-growths that achieve MCG. The universe is essentially popcorn from heat like in a Jiffy popcorn pan building psychology.

The End of Star Manufacturing – What's In Store for Us?

TL;DR: The conclusive marker for the end of anything in the universe is the end of light. Exploring the positive aspects of conscious growth helps in understanding its intended reason and purpose. This gives us humans a lot of time to figure it all out.

Astrophysics and the study of the stars is an extremely important field that has led to many developments in our understanding, ranging from discerning our motion through our solar system, the birth and death of stars, the strange behavior of black holes and even stretching on to help us glean insights into the age of the universe. Astrophysicists often try to determine the life cycle of stars by observing their light as well as the composition of various regions inside galaxies in order to better understand the dynamics at play during star formation. Astrophysicists are also interested in how various stars will end.

Stars eventually use up their fuel (hydrogen, helium and other elements) through nuclear fusion, giving out much of the mass-energy in the form of radiation, which causes them to undergo various transitions in their life cycles. Depending on their mass and composition, some stars will become red giants; others will become white dwarfs after their main sequence of glowing brightly, like our Sun; and some are so small they ignite the fusion process, but remain red due to their lower temperatures. Some of these red dwarfs can contain materials that can also classify them as brown dwarfs. The smaller the mass of a star, the lower the temperature and the longer it will burn. Main sequence stars will burn up some of their hydrogen, which converts to helium during fusion, and begin to fuse helium next. These stars become red giants, which will eventually radiate out their outermost layer, called a planetary nebula, after the helium supply runs out. This ejection leaves behind a white dwarf. As can be seen, various stars will undergo different phases based on their size and composition. However, the end result is that the star cools and will eventually, in some cases over extremely large periods of time, no longer produce light. One scientist has attempted the calculation of how long it will take for black dwarfs, stars that no longer produce thermal radiation, to fully deteriorate and explode.

"Caplan calculates that the most massive black dwarfs will explode first, followed by progressively less massive stars, until there are no more left to go off after about 10^{32000} *years."*[17]

This is such an extremely long period of time that the universe will likely not be recognizable when comparing it to what we see today. The question that remains is not how stars die but how stars continue to form.

Studying the life cycles and distributions of stars throughout the universe helps in gauging the timeline for and the number of conscious-growths that can evolve to their MCG in the universe. The end of star formation is currently thought to occur when the expansion of the universe reaches a point where the matter and elements that make stars begin to cool or dissipate so much so that the temperature required to sustain fusion in a new star is no longer achievable. This also shows evidence that the universe itself has a limited timeline within which life can form in star systems. The universe physically limits the amount of conscious-growth that can result from the Big Bang. Amazingly, the amount of star formation that remains is something that can be estimated. The graph shown below details the density of stars formed per unit time in the universe based on the observations made by Lebanese astrophysicist, David Sobral. Sobral goes on to say that his data strengthens his prediction that the universe is producing its last 5% of stars.

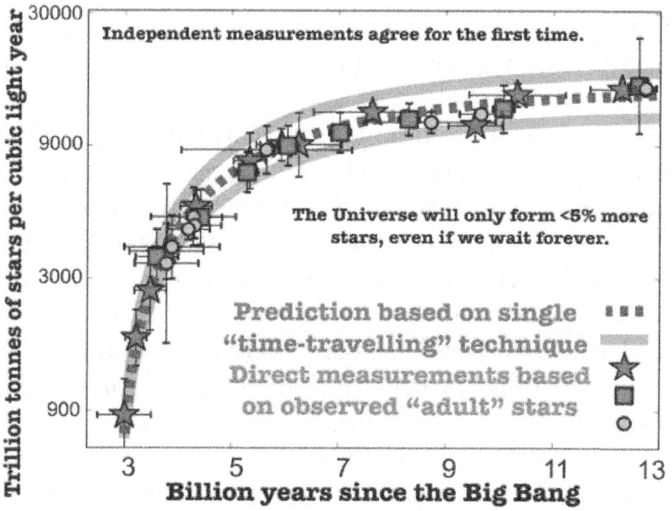

Figure V-671 (Star Formation Timeline)

This figure demonstrates that based on adult stars there are only approximately five percent of stars left to form. (Credit: D. Sobral)

"This diagram shows how the total mass of stars in the Universe should have changed over the last 11 billion years based on the new observations (lines) and how it actually did (symbols; different measurements by other teams). This provides an excellent agreement between both and strengthens the prediction of the new results that no more than a further 5% of stars will come into existence, even if we wait forever."[18]

This comes as evidence that the Earth is on the tail end of the universe's conscious development, as fewer new stars will form. One can also see that the formation of conscious-growths is becoming rarer and rarer through the dynamics of star formation and death. Given that the rate at which new star formation is also decreasing, these last potential conscious-growths can be spread out over vast periods of time. This also decreases the chance that these conscious-growths may come in contact with one another. In addition, while red dwarfs are the most common type of stars in the universe and these stars are so long-lived that they can support the formation of life, they don't in any practical terms enforce the second fundamental limitation of inevitable death with any true sense of urgency. Conscious-growths in these star systems would be less likely to be pushed toward achieving TES and ESS, in comparison to star systems that form with stars like our Sun, because there may not be enough of a stress imparted by the temperature funnel of these star systems to push that conscious-growth into a MCS. Coupling these two facts together, the amount of life that can continue to form is likely even lower than this 5% seems to imply.

BBP The most obvious limitation on the galactic scale is in the eventual sizzling out of stars, which could suggest that conscious life is an intended growth. The death of stars forces consciousness to avoid its demise by needing to exist in that solar system and survive beyond the lifespan of the stars that incubated them into consciousness to begin with. As consciousness grows more aware of its solar death date, it must travel away from its dying star. Technology and genetic adaptability are the only "Noah's Ark" method of avoiding these circumstances. Not only must consciousness escape its planet's demise, but it must also continue out past the solar system to avoid the end of its star's life cycle. As is true with living systems where there is a period of time for birth to occur, followed by growth and then death, the universe operates via similar principles. Looking at the universe over such large cosmic timeframes, it becomes obvious that the Big Bang occurred in order to allow for a phase in which consciousness could be grown and, subsequently, matured. The time for that to be the common modality in the universe is and has been ending, or at least the number of consciousnesses formed is significantly decreasing. These phases demonstrate that, like all other forms of life, the universe is itself spawning life and was created to do so. It follows the same dynamics as other living systems. As the stars begin to fade out, inevitable death forces conscious-growths to survive the end of the luminous era in preparation for the UGCH. From bang to burnout, the universe is ideally crafted to spawn conscious-growths and push them to their maximum potential. It is important for a conscious-growth to be unaware of the end of the universe, which is consistent with HTO. For if this was not the case, a determination of origin could be made and that determination could harm the continued growth of consciousness. Therefore, this theory contends that the universe must continue to expand and never contract, as this would allow a calculable origin point of the entire cosmos.

Potential Ends of the Universe

TL;DR: If we are a fungal conscious-growth, then it would make perfect sense that no matter how the universe or growth of consciousness ends, that there would somehow be a harvesting of such a well-intended growth, which we will term the Uni-Galactic Conscious Harvest (UGCH).

The theories and mechanisms discussed prior are all part of a grand framework that describes how the universe evolves and cultivates consciousness through HTO. However, this theory does not describe what happens as the universe expands for extremely long periods of time. The universe may die in a whimper; it may start anew from another BB; or it may even continue to expand indefinitely. This next section is devoted to discussing possible outcomes for the universe as it continues to expand over extremely long periods of time. The reader is left to decide which one of these possible outcomes, or some other outcome not yet discussed, most resonates with the material presented thus far.

Complete Burnout due to Expansion (The Big Rip)

One potential outcome for the universe is that it will continue to expand indefinitely. From what we can currently see, the universe continues to expand at an accelerated rate. That is to say the very fabric of spacetime is expanding. Objects that are very far away from us are even receding faster than light! This does not violate any principles of physics as it is not the matter and objects within space that are accelerating faster than light but the fabric within which these massive objects sit.

Should this expansion continue in this manner, eventually galaxies will be so far apart from each other that no other stars from their galaxies will ever be seen. The light from distant galaxies will travel toward others at the speed of light. However, since those galaxies are receding faster than light, it will be impossible for that light to arrive anywhere else. Ultimately, it will be galaxies resting in a seemingly endless void.

It is at this point that two possibilities can occur. The first possibility is that this expansion continues further and begins to rip galaxies apart and then the fabric of spacetime itself. This event has been theorized and is called the "Big Rip." Many things occur before the fabric of spacetime is ripped apart, such as solar systems, planets, molecules and atoms being torn apart first. However, the end is inevitable and can be described best as nothing left in any form of condensed matter exists. This would be the point at which the universe returns to pure energy. Interestingly, black holes can still exist in this new universal state but their topological interiors change such that as they merge, inflation of spacetime can begin. This could potentially be the start of another universal Big Bang expansion. It could also be something uniquely different.

The other option is that somewhere before the "Big Rip", a civilization becomes advanced enough to master the principles of quantum mechanics and quantum gravity and spawns another universe of their own. In essence, this civilization creates another universe in order to escape inevitable death. Whether or not they are able to exist in this new universe or whether they simply set initial conditions in motion in order to promote the formation of consciousness in the next universe is entirely speculative. However, in this situation, the seeds of life are set in motion and play out over billions of years in a manner similar to what has occurred so far in this universe.

Engulfing through Black Holes and Black Hole Mergers (Photon/Graviton Soup Model for CCC)

It is also possible that the universe never enters a state of increasing expansion that would generate a "Big Rip." In this case, the universe can continue to expand indefinitely without the destruction/destabilization of matter. Once again, two scenarios are possible. In one scenario, galaxies will recede so far away from one another that every galaxy is essentially surrounded by nothing but void. This is similar to the situation of the "Big Rip," though, the universe dies out in a slow fizzle. Stars eventually burn out and no other stars continue to form. Light becomes dimmer and dimmer until no light is emitted at all. The universe continues to cool and expand, dying in an unremarkable death.

The other possibility is that during this expansion, all matter eventually is sucked into black holes. The star systems and nebulae surrounding a galaxy's central black hole all eventually succumb to its gravity in an inward spiral. Once all matter has been sucked into black holes, the energy is radiated away over extremely long periods of time through Hawking Radiation. The energy released is in the form of gravitons and photons, which are massless. Since these particles are massless, they experience no time or space. Ultimately, at this point, it could be considered that the universe looks a lot like it did in the early moments of the Big Bang. This theory of the universe is called Conformal Cyclic Cosmology (CCC). Basically, since no matter is left to experience space or time, a rescaling can occur such that the very vast and wide universe is simply akin (mathematically) to the tiny point at which a bang could occur once again. This theory was proposed by famous astrophysicist Roger Penrose.

In this particular end to the universe, it is possible that a sufficiently advanced civilization could somehow protect themselves from the wrath of the black holes guzzling matter inward as well as the reigniting of the next Big Bang. In this case, the consciousness that had matured in one universal cycle may go on to evolve throughout the next. There is some evidence that our universe is not the only one that has existed in the structure of the CMB. A study, also led by Roger Penrose, suggests that black holes from prior universes have influenced the structure of our current universe and that this influence can be seen in large "cold" spots found in the CMB we see today.

"A team led by Oxford University mathematics emeritus Roger Penrose, a former collaborator of the late Stephen Hawking, claims in a new paper lodged on the preprint server arXiv to have found signs of so-called Hawking Points, anomalous high energy features in the CMB…' This burst of energy from a now decayed black hole then spreads out quickly in our newly formed universe, leaving a warm central point with a cooling spot around it,' says astronomer Alan Duffy from Australia's Swinburne University and Lead Scientist of the Royal Institution of Australia, who was not involved in the research. 'In other words, they have proposed that we can search for an echo of a previous universe in the CMB'."[19]

The Big Crunch

The theory of the Big Crunch was widely discredited for quite some time, given the fact that the universe seems to be expanding at an accelerated rate. However, recent research suggests that dark matter may actually be better described as a quantity called quintessence. This property describes how the strength and thus effect of what we now call dark energy can change throughout the life of the universe. In a recent study, scientists were able to tweak their model of quintessence such that the behavior of this changing quantity was able to accurately represent the universe's history up until now. They then used the model to peer forward and see how the universe would behave under this new assumption.

"To their surprise, dark energy in their model can decay with time… Its strength can weaken. And if it does so in a certain way, then eventually the antigravitational property of dark energy goes away and it transitions back into something that's more like ordinary matter."[20]

Interestingly enough, this was proposed to occur relatively quickly in terms of the universe's total evolution so far. The authors stated that the slowing of the universe's expansion could begin as quickly as 65 million years from now, which cosmologically speaking is quite short. The resulting crunch would take many billions of years after this point in time.

Summary

Some of these theories seem to suggest a universe that would be consistent with the three fundamental limitations and preserve HTO and produce MCG. The others leave a little wiggle room for interpretation as to what may come next and do not mesh as well with the fundamental limitations. In the case of the Big Crunch, it is quite clear that the universe is forcing consciousness to reach an ultimate form by way of enforcing the second fundamental limitation of inevitable death. This is also the case with the theory of the Big Rip, with the caveat mentioned before of another universe being instantiated if some civilization were to have reached a point of evolution in which advanced enough technology would allow for surviving the Big Rip in this manner. However, in the case of CCC, there is potential room for escape into a completely unknown phase of the universe, after the universe undergoes a conformal rescaling as time and space have no meaning to the remaining massless particles of gravitons and photons. This particular theory seems to allow for the chance that things can start over, possibly improved by the prior iteration somehow.

In accordance with the three fundamental limitations, certain outcomes for the end of the universe have been demonstrated to be logically consistent. Others did not play out in a way that is consistent with these principles. By observing which outcomes are compatible with the fundamental limitations, a clearer picture of the universe's end can be achieved.

Additionally, a purpose can be ascribed to that end through understanding the means by which it came about. This chapter will give you a functional understanding of the purpose of the universe and how it works and strives to achieve the goal of maximizing consciousness.

2. BIOLOGY

Introduction

The perplexing question of life and consciousness in the universe has intrigued humanity for ages, from cave drawings depicting divine or extraterrestrial influence to the musings of early philosophers. However, we can find answers by observing the universe's production. The evolution of organic life, starting from a single cell and progressing through various forms, culminates in the creation of the Tin Can Man (TCM), symbolizing Eternal Self-Sustainability (ESS). Consciousness development adheres to the guidelines of HTO, operating slowly and expansively. The branching nature of life, comparable to fungal growth, extends like mycelium underground. This behavior is termed fungal, not literally discussing fungi. To understand humanity's appearance and conscious growth, it's beneficial to comprehend the properties and life cycles of fungi. This chapter emphasizes that all life is essentially a *fungus*, denoted in italics, and explores how fungi grow to produce human spores capable of continuing existence far beyond their home planet.

What is Fungus?

TL;DR: Scientists today see fungus as an alien life form because it behaves more like an animal than a plant. It also needs no sunlight. Ancient fungus was actually both plant and animal long before today's evolution of plants or animals. About one billion years ago, the plant and animal branches distinctly split away from fungus (The Split). Today fungus has evolved to a kingdom all its own, acting as a decoy and hiding its true and ancient purpose. It appears to be a growth method that holds evidence of a creator intending to grow a consciousness with no strings attached through the concept of untraceable or Hidden True Origin (HTO), which could be evidence of System (EOS).

Fungi are in a kingdom of its own today, but this was not the case earlier in biological history. The earliest known forms of life were the eubacteria and archaebacteria. Over time, these first single-celled life forms evolved together to become what are known as the protista. The eubacteria evolved along with protista to form the kingdom plantae, which contains the various trees, vegetables, fruits and other plant-like structures we know today. Other evolutions from protista formed animalia, consisting of animals and fungi, which today contain some elements of the animal world as fungi split from both plants and animals sometime very far back in the evolutionary progression of life on Earth.

"As it turns out, animals and fungi share a common ancestor and branched away from plants at some point about 1.1 billion years ago. It was only later that animals and fungi separated on the genealogical tree of life, making mushrooms more closely related to humans than plants. Most likely, this common ancestor was a single-celled organism that had sperm-like characteristics (like an animal) and then a later developmental stage with a stronger cell wall (fungi)."[21]

Fungi consist of a group of spore-producing multicellular organisms feeding on organic matter, including molds, yeast (which happens to be unicellular), mushrooms and toadstools. They possess the capability to break down organic matter, which makes fungi the ideal candidate for transforming biospheres through their ability to process a large variety of compounds. This allows for a much more rapid transformation (terraforming) of a biosphere than is possible by normal geological processes or other life forms. Fungi, in algal form, are thought by scientists to have been one of the major contributors to the production of oxygen, which forms one of the most important components of our atmosphere. Some fungal forms have the ability to move from place to place, unlike bacteria, through the use of spores that eject from the lamella (underside of the umbrella-like structure of a mushroom) of the fruiting body of a fungus. It does this by generating its own burst of air that sends spores floating through the air, allowing for reproduction and expansion of the organism. Fungus is critical for the massive transformation of nutrients that make life possible on Earth.

Giant Mushrooms First Cover Earth

TL;DR: Here is an example of how the fungus used the fruiting body of a mushroom to first spread over land. This breaks down rock and sand into soil along with the dead mushrooms that made way for the fern revolution, allowing for the insect revolution. That process of planetary development is very fungal in its approach to prepare and produce the line of food needed for what would eventually be the amphibious revolution. And later, of course, the rise of oxygen for the bipedal brain and its conscious capacity.

Long before trees, the Earth was covered with large mushrooms. It may be that there was not enough solar exposure for plant life and being that fungus is the starter of cellular life, it grows out of control. This would have also started the decaying and feeding process needed for plant life to later thrive.

'From around 420 to 350 million years ago, when land plants were still the relatively new kids on the evolutionary block and 'the tallest trees stood just a few feet high,' giant spires of life poked from the Earth. 'The ancient organism boasted trunks up to 24 feet (8 meters) high and as wide as three feet (one meter),' said National Geographic in 2007. With the help of a fossil dug up in Saudi Arabia, scientists finally figured out what the giant creature was: a fungus. (We think.)'[22]

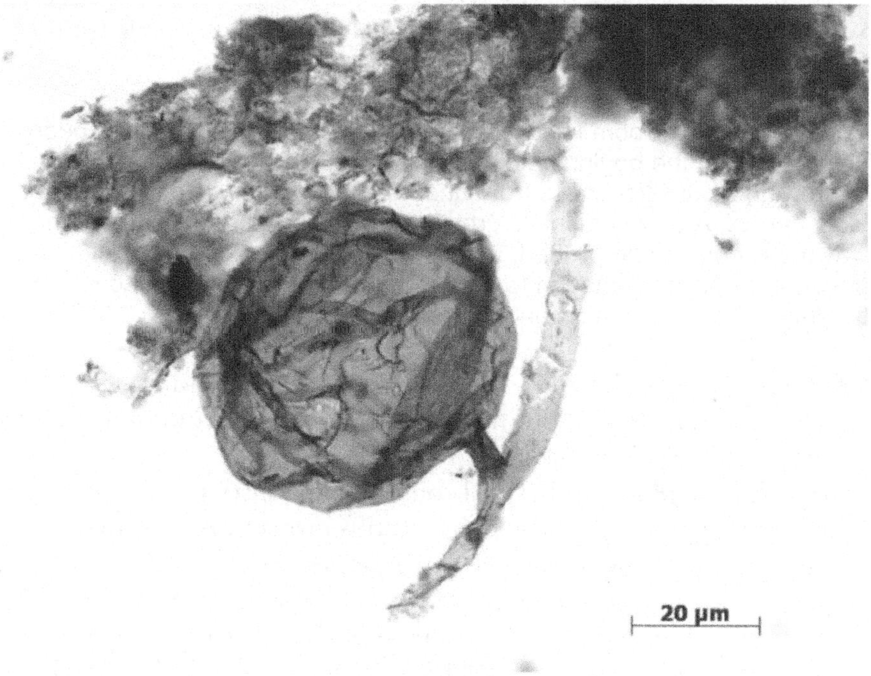

Figure B 2.3 | Fossilized Fungus

This is a photo of a fossil of fungus (Ourasphaira giraldae ~1.01-0.89 BYA) discovered preserved in shale of the Grassy Bay Formation in Arctic Canada. This fossil finding predates the earliest fossils found so far that clearly represent fungal character by around half a billion years. This provides compelling evidence that it is perhaps fungus that started all complex multicellular life on Earth and helps with the understanding that all complex multicellular life is still some variation of this organism.

EOS

- Starting with the smallest of cellular development
- Not dependent on sunlight
- Splits from plant and animal
- Oldest fossils
- Responsible for the foundation for all plant life and forestry
- Capable of memory and communication
- First single cell then multicellular life
- Foundation for the growth of animals, aquatic, land and sky
- Forms the cultures all plants and animals rely on today

Culmination of Organic Life: Comparison with a Mushroom Life Cycle

TL;DR: The mushroom life cycle is the best example of what has been going on since the beginning of it all. This cycle represents the best method of growing a consciousness from undetectable origins galactically, but not biospherical. A consciousness growing in this manner seems to be the only way it could be accomplished, by obscuring the past long before it becomes conscious enough to recognize it and then forcing it to respond to the future death of its birthplace.

Comparing the mushroom's life cycle to that of the entirety of the development and evolution of organic life (starting from multiple forms of simpler life, such as algae and bacteria, all the way on to lions, tigers and bears...and what lies beyond--oh my!) demonstrates the possibility that humanity is the pinnacle phase of this growth process, much like a mushroom preparing to achieve spore release. With more advancements in technology, when humanity journeys away from Earth for the final time, it will be akin to the ejected spore of a mushroom. From this point forward, we will look at humanity and all preceding life as one single fungus, in order to further motivate this theory and comparison. Use of the non-italicized version of "fungus" will be exclusive to the literal fungus that has existed and currently exists on Earth today.

In order to understand the comparison between a mushroom/fungus life cycle and the timeline of events that have occurred throughout the history of the evolution of life on Earth, it is instructive to draw out and detail the parallels between the mushroom/fungus life cycle and the various periods that the Earth has gone through up until the modern day and beyond. In order to clearly illustrate these associations, Figure 222 has been created to aid the reader in understanding this comparison.

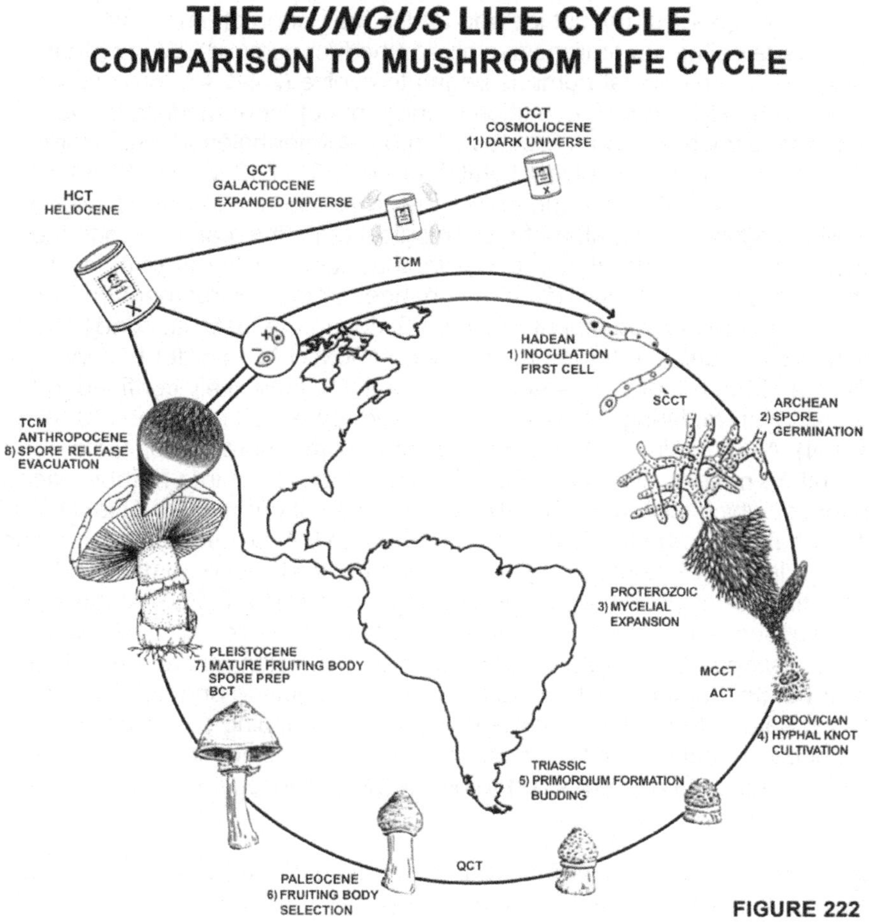

Figure 222 (Fungal life cycle comparison)

This figure compares the evolution of humans to the life cycle of a mushroom. (1) Inoculation–the Hadean eon is the earliest years of the Earth where the seed material for what will eventually become life is all that exists on the planet (4.5 BYA - 4 BYA). (2) Germination–the Archean eon is when the earliest life, the Last Universal Common Ancestor (LUCA), began to form and evolve. This is the beginning of the SCCT (4 BYA). (3) Mycelial Expansion–the Proterozoic eon is one of the largest in geologic time where roots, plankton, and other forms of algae begin to develop, as well as some forms of underwater plants (2.5 BYA - 0.5BYA). (4) Hyphal knot–the Ordovician period, which occurs just after the Cambrian period where aquatic life (the ACT) evolved, is the period in which plants and insects began to evolve (0.49-0.44 BYA). (5) Primordial Formation–the Triassic period marks the rise of the reptiles, coming after the evolution of the QCT, including the dinosaur (0.25-0.20 BYA). (6) Fruiting body–the Paleocene marks the period of time in which primates evolve, living alongside other

QCTs. (66-56 MYA). (7) Mature Fruiting body–the Pleistocene period encompasses the last ice age and is the period of time in which Australopithecus afarensis evolved (the first hominid). With further evolution, the earliest humans began to evolve (2.4-1.4 MYA) and was Homo habilis. Further evolution brought about more modern forms of humans such as Homo erectus, Homo Neanderthalensis, and Homo Sapiens began to evolve. (Total timeline: 4.0-0.012 MYA). (8) Spore Release–the Anthropocene marks the current period in Earth's history. Mankind is now capable of technological feats that can influence the environment on Earth. It is at this point that humanity is transitioning into a new phase of existence. (8+) The changes to the environment on Earth promote the development of the GCT due to the fact that conditions become inhospitable for life to continue ~100MYFN. In order to survive, a Tin Can Man will exist in space permanently but still likely confined to the Solar System during this period, proposed to be the Heliocene (100-200 MYFN). As the solar system continues to become more inhospitable due to the nature of solar dynamics, the TCMs must move out into further reaches of space, which include both within and outside of our galaxy. This period is hypothesized to be the Galactiocene ~300 BYFN. Once the universe has expanded for extremely long periods of time, it will become dark and can essentially be considered in a "totally expanded state" (TEU). At this period and subsequent periods, continued expansion can occur but will likely not be distinguishable from other periods of expansion due to the lack of information that can be transmitted from one region to another. This is the case due to the fact that the universe is expanding so rapidly that electromagnetic signals cannot travel from one region to another. This period is hypothesized to be the Cosmoliocene ~1 TYFN.

Extending this concept, one can look at the general behavior of humans and relate them to the spore in a mushroom life cycle. Humans have gathered and collected materials, transforming them, similar to what has been discussed about fungi scavenging their surroundings to sequester minerals from rock. Humans also have similar behaviors when it comes to influencing plant life. Due to our farming activities, and our attempts to produce better (or ever-tastier) varieties of plant life, humans have created niches of vegetation that would not have existed otherwise. This can be likened to the mycorrhizal networks formed between the roots of plants and the mycelia of fungi. The nutrient exchange that both participate in is mutually beneficial. The spores of mushrooms that grow and spread out from the fruiting body can be likened to our technological expansion in urban landscapes. Each of these behaviors is directly comparable to that of fungus.

Looking one step beyond, these fungal spores can be ejected up to hundreds of feet. At the scale of the tiny fungus, this is a moonshot to say the least. Humans have also begun to explore our local galactic neighborhood, with the ultimate intent to colonize the cosmos. The human spore is nothing more than a vehicle for *fungus* to escape once TES (Technological Evacuation Standards) has been achieved.

BP This comparison to the mushroom cycle is the best example that can be given of the deterministic internal logic of evolution, which dictates how events must play out on a biosphere in order to reach TES and a MCS. The growth process of the *fungus*, with starvation/predation motivating adaptation and intercellular communication facilitating improved response times to the surrounding environment, is necessary for consciousness to have evolved. This is the case because without these factors at play there is no impetus to make changes or improvements in the way in which nutrients are gathered and assimilated or in the ability to escape predators. By the time the *fungus* has produced a BCT (man) and subsequently reached the spore-release phase of the mushroom cycle, it will have made leaps and bounds in technology that allow the continuation of the *fungus* in space and on other planets. The *fungus* always develops through each and every event, all being necessary, to survive Earth's end. This means that every trial and tribulation, no matter how evil or just, is an absolute requirement in order for the *fungus* to be successful in achieving TES.

EOS

- Starts early in the development of a biosphere
- Evolves from water to land then spore
- Eventually develops technological capability of leaving its birth biosphere

What is The *Fungus*?

TL;DR: All organic life from the first single cell to the last cell to leave the earth is one large and very old *fungus* developing the growth of a consciousness. All life on Earth is for the success of the spore of that *fungus*, Humanity. There are two phases of any *fungus* that may form on other planets in the universe, terrestrial and post-terrestrial fungus. The behavior of all plant and animal life on Earth (or any other biosphere throughout the universe) is a self-feeding cannibalistic *fungus*, which is evolutionarily striving for the technological ability to survive the end of its biosphere. Andrew Fiouzi cites James Gallagher on human cellular composition and the share of microbes in the same:

"If God created man in his own image, then that means God is one filthy bastard. That's according to a study reported by the BBC last month, which states that more than half of the human body is made up of microscopic colonists and not, y'know — human stuff. "Human cells make up only 43 percent of the body's total cell count."[23]

That is saying the rest of the human stuff, 57 percent, "are cells that are fungi."

Because of fungi's ancestry and what it has evolved into today, it is possible that all complex multicellular life started from fungus (fungus likely being the first multicellular life that evolved much earlier than complex forms) and has grown into all of the various forms of complex life we see today; making all plant and animal life one single larger fungus or Plant And Animal Fungus (PAAF).

This is arguably the case because the fungus (PAAF) that began evolving billions of years ago gave rise to both plant and animal life, such that it first split away from the plant kingdom and then the animal kingdom not long after. This splitting of fungus into other kingdoms of life was necessary as it acted as a balancing mechanism in a system with a limited supply of nutrients. This split allowed the plant kingdom enough time to evolve its nutrient gathering/producing capabilities in order to generate what would later become vegetables. This evolutionary process is absolutely essential because the increased nutrient density in vegetables is required in order to support the energy requirements of brain growth in future animals and ultimately, the spore of the *fungus*. Because these splits took place so long ago, and fungi represent their own unique kingdom, it increases the difficulty in tracing back the origins of life. If the world only evolved the two simplest forms of bacteria (eubacteria and archaebacteria), eventually all of the nutrients that these organisms fed upon would be used up and life on Earth would have stagnated or died. However, the splitting of plant and animal from fungus allows a system in which animal consumes both plant and fungus; the waste of animals fertilizes the plants, and the dying plants and animals are decomposed by the fungi. This generates a self-sustaining "cannibalistic" ecosystem in which the *fungus* feeds upon itself in order to survive, grow larger, and become more aware.

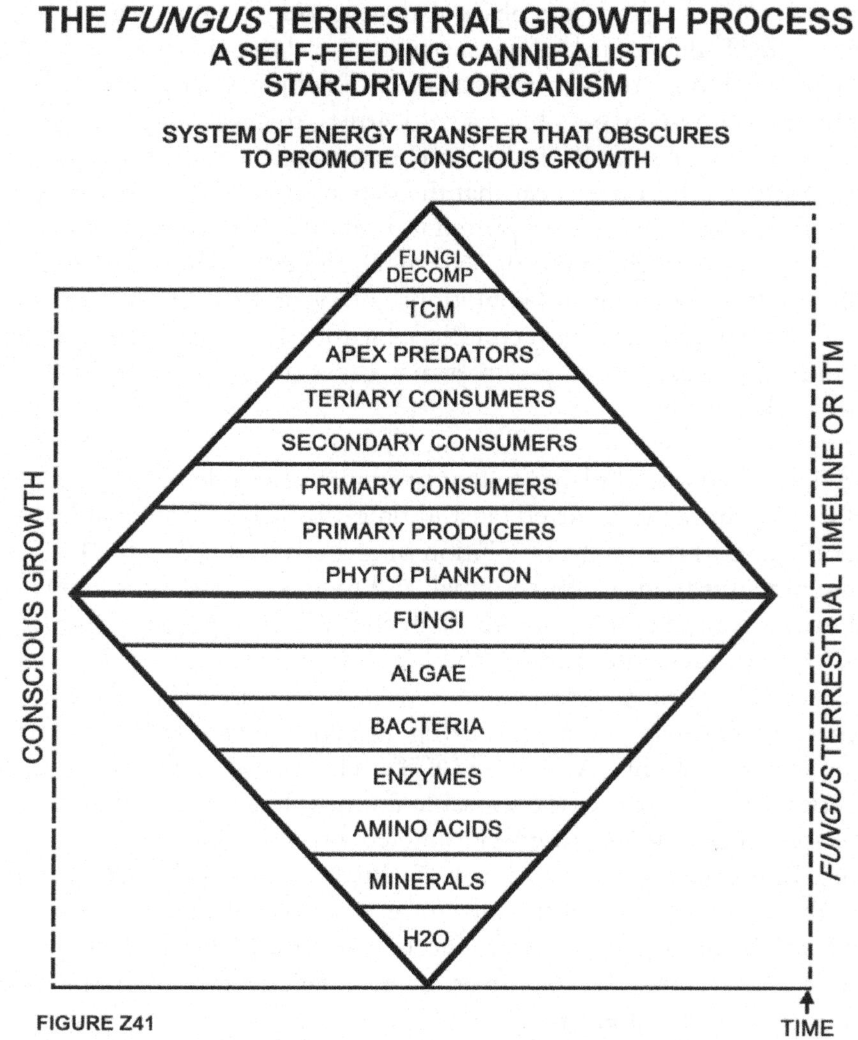

Figure Z41 (Star-driven self feeding)

This figure shows the sources of self feeding predator prey from the beginning to the end of Earth's organic occupation. The *fungus* first needs building blocks that make up the foundation to evolve to a single cell. It shows how the self-feeding evolutionary process is structured. It is a very strict system only allowed to evolve from sunlight. It is through this process that a consciousness grows to maturity and leaves its birth biosphere.

We are totally dependent on all life forms, since we are all part of the fungus, until we develop sufficient self-sustainable technologies and artificial replacements for what we currently obtain from the self-feeding cycle inherent in the fungus. Given the difficulty of these extremely technical feats, it is obvious that the only creature capable of them is a BCT, due to its biological structure and the increased level of consciousness this structure imparts. As an example, the opposable thumb opens up many possibilities for learning that are not afforded to other organisms. As the BCTs become increasingly aware of their terrestrial fate, it becomes clear that the dependency of this organism on the food chain that has supported it until this point is problematic. This is the point at which the BCTs may be forced underground as a result of inhospitable surface conditions and begin their preparation to become TCMs and develop technologies with the goal of ESS. In other instances, the depletion of resources may cause the *fungus* to evacuate its terrestrial birthplace.

ITM The evolution of plant life that produces vegetables is providential in that it occurred "just in time", by geologic standards, in order to support the further evolution of animals to develop the BCT. By the time this triad of plant, animal, and fungus had evolved, all life on Earth was primed for going through all the evolutionary iterations required to produce the spore of the *fungus*. It takes millions to billions of years to allow the *fungus* to iterate to a "solution" that is intelligent, adaptable, and capable of manipulating its environment. The transition from prior animals, like ACTs and QCTs, allowed for increased dexterity through the use of opposable thumbs. It is at this point that the *fungus* is truly capable of developing technologies that will be required to survive the end of Earth's habitable period. The *fungus*' goal, from the very beginning until evacuation, is to formulate and escape. It starts with being able to move around in general, whether in water or on land, then evolves to move from water to land, and later from land to sky, then finally outer space.

IPT Eventually, when the awareness of BCTs of Earth's *fungus* reaches a point where it becomes clear that its demise is impending and inevitable, due to the destruction of its entire biosphere, it begins to develop its spore. Humanity is in its beginning stages of transitioning into this spore through advancements in technology and our increasing awareness of the eventuality we are faced with.

BP Comparing the timeline of organic life's development on Earth with all of the separations in biological kingdoms discussed before, to the mushroom life cycle in Figure 222, it becomes easier to see that all life on Earth is one large organism constantly developing to some final end goal. From single-celled organisms to the TCM, this large organism displays properties that are analogous to fungi, which makes it quite appropriate to call this organism *fungus*. Just as the fungi split from plant and animals in order to ensure sustainable growth through a self-feeding cycle, any other impedance to the growth of the *fungus* will be met by some form of adaptation of the *fungus* to iterate closer toward achieving TES. Growing the human organism via the *fungus*, with all the various splitting and branching of one organism to another along the way, is substantial evidence that HTO is the method by which the universe farms consciousness for MCG.

BBP Imagine each biosphere in the universe like a giant petri dish, only spherical. With the correct temperature and time allotted, a biological culture begins. That culture starts much in the same way a culture is made by laboratory scientists in a petri dish today. They are both cultures, grown in temperature-controlled environments and by external heating sources. If life can sustain long enough anywhere, it should evolve into a self-feeding plant/animal system. Eventually, intelligent conscious-growths will emerge as a product.

These conscious-growths will be pushed to maturity through the ultimate stressor: the temperature funnel. Once these conscious-growths reach a MCS, the *fungus* that has developed on that biosphere will have produced a spore and will begin to survive out in the cosmos. This process happens many times over on different biospheres at different times. Some fungal growths may interact with each other, whereas others may have existed so far away or separated in time from other fungal growths that they never interact. Ultimately, the goal is to produce MCG for UGCH.

EOS

- A biosphere with H_2O and minerals
- Starts cell division
- Develops a consciousness
- Evolves to bipedalism
- Develops technology
- Escapes its own biosphere before annihilation.

This method hides the origin of what will become capable of finding it (also called Hidden True Origin).

Fungal Properties

TL;DR: It is and has always been about communication and its speed. Gathering information and transferring it is equal to the rate of that *fungus*' conscious-growth. If it is telling where the food is to project for safety and story. The *fungus* is always slowly adapting to perfect its communication for terrestrial success by eventually becoming nonverbal in the future.

In order to understand what it means for all life to be one *fungus*, it is instructive to demonstrate properties that fungi possess and, thus, all subsequent life should possess. These properties will be based on various aspects and behaviors of the growth of *fungus*, the way in which they communicate and the way in which they can spread over vast distances to support multiple other organisms and ecosystems. Separately, each of these facts about fungi are awe-inspiring and demonstrate their pivotal role in the history of our evolution.

Pervasive and Supportive

ITM Specific species of fungi on Earth today are descendants of the oldest multicellular organisms. These fungi are the foundation of all the forestry that produces the oxygen required for animals to breathe by supplying the plants within forests with increased levels of nutrients. This is an important facet of fungal behavior and will be discussed further in detail later. However, of intrigue is the fact that the largest land organism on Earth today is a fungus.

IPT A giant Armillaria ostoyae that was discovered in 1998 set a new record as the world's largest known organism and is evidence that the foundation of all life starts and continues because of the pivotal role that fungus plays in balancing and supporting our ecosystems.

"Next time you purchase white button mushrooms at the grocery store, just remember, they may be cute and bite-size but they have a relative out west that occupies some 2,384 acres (965 hectares) of soil in Oregon's Blue Mountains. Put another way, this humongous fungus would encompass 1,665 football fields, or nearly four square miles…"[24]

BP Plant and animal life would never have evolved on Earth without fungi to support them. To this day, they still require a variety of fungal organisms (like Armillaria ostoyae) living underground that support them. This fungus is both the foundation of other primary producers as well as a primary producer itself. It is also a secondary consumer, as fungus feeds on decaying organic matter (animal, plant, and insect). This versatility allows for fungus to be a solid foundation for all life that comes thereafter. Fungus will continue to support all life on Earth until life can no longer be sustained at all. When humans are incapable of surviving on Earth's surface, they will be forced to join their fungal relatives underground because of the ramifications of the temp funnel. This fortifies the fact that even though some separation in kingdoms has occurred, all life shares the common ancestor of fungus and is continually supported by that *fungus* to the very end.

BBP In order to support life in any biosphere, a widespread organism that is symbiotic with various forms of life by both producing and consuming nutrients is a necessity. After realizing the scale of *fungus*, it is nearly impossible to lose sight of the fact that it serves this purpose, not only as the igniter of complex organic life but the basis for growing a collective consciousness. This organic mechanism is the one that is most likely to produce complex multicellular organisms with an ever-increasing consciousness throughout the universe. This is because of the basic principles that are thrust upon systems by the temperature funnel and the timelines for cellular division. Thus, it can be expected that each biosphere that reaches sufficient complexity to begin generating conscious-growths should house an "organism" much like the fungus we see on Earth. This *fungus* supports the growth of collective consciousness until a "spore", or the pinnacle species of the biosphere, is produced. In doing so, the *fungus* continues to support conscious-growths on their journey toward TES. Throughout the universe, the success or failure of any *fungus* on a biosphere in producing a conscious-growth that achieves TES is random, which further supports the fact that the universe operates via HTO in order to produce MCG.

Fungal Networks and Communication (Similarities to Human Communication)

TL;DR: It would seem that you can't make an intelligent conscious entity just appear out of nowhere; it would be more likely that for it to be truly deserving of its true intelligence that particular consciousness would have to be grown in an environment that hides its true origin (TO) at the same time. To do so it would literally have to be grown from fungus itself.

ITM Fungi, like modern humans, also utilize complex network structures to communicate. Similar to the vast array of wires that are both above and below ground to transmit electrical signals, fungi utilize filamentous structures that propagate underground in order to transmit information to different parts of the organism. These networks of filaments are called mycorrhizal networks.

"Fungus expert Paul Stamets called them 'Earth's natural internet' in a 2008 TED talk. He first had the idea in the 1970s when he was studying fungi using an electron microscope. Stamets noticed similarities between mycelia and ARPANET, the US Department of Defense's early version of the internet."[25]

This demonstrates that as fungi have evolved from their earliest forms, which were mainly single-celled organisms, they have created structures that allow for a wide variety of communication. Interestingly, this communication occurs not only between parts of the fungus but also between other species entirely.

"These plants are not really individuals in the sense that Darwin thought they were individuals competing for survival of the fittest…In fact they are interacting with each other, trying to help each other survive. By linking to the fungal network they can help out their neighbours by sharing nutrients and information – or sabotage unwelcome plants by spreading toxic chemicals through the network. This 'wood wide web', it turns out, even has its own version of cybercrime."[26]

Here we have evidence that fungi are doing exactly what has been described, which is, supporting other organic life. One way in which they are able to directly support other forms of life is by increasing their available area of absorption. These mycorrhizal fungi increase the surface absorbing area of roots 10 to 1,000 times, thereby greatly improving the ability of the plants to use the nutrients and water.

Consciousness can be demonstrated by the fact that invaders or hostile entities are warded off while friendly neighbors are supported. In order to be successful in assisting the growth of plant life, an inherent understanding of the various members in a fungal-root network would be required. This demonstrates some of the earliest forms of conscious awareness that evolved on Earth. These conscious organisms were not only aware of themselves and their surroundings but also of the needs of other species and could determine which was friend or foe. This type of awareness across species is exactly the type of evidence that suggests that fungi both support all forms of life that come subsequent to it and are a necessary component for this consciousness to continue to grow. These fungal networks are found all across the globe and, as mentioned prior, can reach truly enormous sizes.

"'Myco' – 'rhiza' literally means 'fungus' – 'root' and describes the mutually beneficial relationship between the plant and root fungus... More than 95 percent of terrestrial plant species form a symbiotic relationship with beneficial mycorrhizal fungi, and have evolved this symbiotic relationship over the past several hundred million years."[27]

These symbiotic partnerships allow for enhanced adaptation to the environment and allow plants to move from swamps/oceanic environments to dryer land in order to span the globe.

"These fungi predate the evolution of terrestrial plants, and it was the partnership with mycorrhizal fungi that allowed plants to begin to colonize dry land and create life on Earth as we know it."[28]

Given that 95 percent of plant life seems to communicate and form a symbiotic relationship between a fungal network and its root systems, it would be no stretch to conclude that fungi have also supported many other forms of life as well. It is then logical to conclude that fungi support animal life on Earth as well, since nutrient availability for herbivores would be seriously diminished without their support.

IPT More recently, scientists are trying to understand where and how consciousness like our own may have begun in our evolutionary history. In a truly fascinating experiment, scientists were able to provide a particular slime mold, a protist (P. polycephalum), with a "puzzle" to solve. Interestingly, the slime mold "solved" it with a better solution than the one that humans have created. The puzzle was an organization of oats distributed in the same arrangement (however reduced to scale) as the subway stations in Tokyo.

"Nakagaki recreated the Tokyo subway, with the station nodes marked out with oats; P. polycephalum recreated it almost exactly – except the slime mold version was more robust to damage, wherein if a link got severed, the rest of the network could carry on."[29]

Other discoveries by teams of scientists were made that demonstrated this slime mold could form memories about the locations of food, or even remember the likes/dislikes to various substances it had encountered previously. Not only were these memories able to be stored, they could be transferred to other parts of the slime mold.

"...researchers found that P. polycephalum can 'remember' where it has previously found food based on the structure of the veins in that area. This followed previous research from Dussutour and her colleagues, who discovered that blobs of slime mold could learn and remember substances that they didn't like, and communicate that information to other blobs of slime mold once they fused."[80]

Even more interesting, the slime mold is capable of solving exponentially complex problems that we test computers' function with. *"Yet another team of researchers found that the protist could efficiently solve the traveling salesman problem, an exponentially complex mathematical task that programmers routinely use to test algorithms."*[81]

Scientists are interested to see the effect of a zero-gravity environment on the performance and abilities of this slime mold. This slime mold will be launched into space for continued experimentation in order to further understand the way it operates.

BP As Earth began to develop life, most of these life forms were content to absorb nutrients from their surroundings and excrete waste. These simple single-celled organisms were not in a homeostatic balance with their environments. Very shortly thereafter, life forms that exhibit a greater degree of consciousness evolved starting with protista, some of which are fungus-like, and evolving toward fungi. These fungi and their fungus-like protista ancestors were capable of much more intricate communication and information processing/storage. These organisms were capable of maintaining balance between themselves and those that they formed networks with. This is an essential property for the development and continual evolution of consciousness. Without fungi, the entire ecosystem that has formed and will continue to evolve on Earth would not be possible. We can see that in the time since bipedal creatures have evolved, technologies and ways to store information have evolved in a way that resembles these fungal networks. It is then obvious that these very skills that were developed early on in our evolutionary history are a necessity that continues to bear fruit and will do so until the point at which the *fungus* is capable of achieving TES.

BBP Across the cosmos, there are innumerable star systems in which life could take root. The earliest life forms will be very simplistic and merely reacting to their environment. However, as evolution progresses, it will become pivotal for an organism to communicate both more broadly and more quickly. Forming network structures will play a large role in both communication and nutrient assimilation. Thus, around the cosmos, an organism much like fungi will be a necessity to begin the transformation from single-celled to multicellular life and maximize the potential for an evolved consciousness. As life continues to evolve, communication speed and distance increase, as does interconnectedness. Though the evolution of language and its properties is discussed further in chapter four of this book, it is clear at this point that fungi form the very foundation of all subsequent complex life.

Each planet capable of hosting life within a star system is in a race to reach sufficient complexity to produce life that resembles a fungus and then goes on to produce a self-sustaining/self-feeding ecosystem. As this ecosystem evolves, many different animals will be produced, capable of communicating in more complex ways and creating more complex network structures. Ultimately, a BCT will develop that is capable of such advanced communication, networking, and manipulation of its surroundings that technology begins to develop. Through the use of technology, the advancements that are required to ensure that this *fungus* achieves TES begin to develop.

At some point, the amount of information that is to be conveyed by these organisms will be too massive to deliver via spoken language. Eventually, telepathic capabilities will arise in which communication can span over large distances and convey complex emotions and thoughts without the need for sequential verbal delivery. This is in line with the fact that the *fungus* continues to improve its ability to both network and increase the speed at which it is able to communicate and transmit vital information. If the biosphere remains sustainable, all that is learned by the *fungus* is recorded and used to continually build upon in order to generate technology that ensures the survival of the *fungus* off-world. It is at this stage that the *fungus* will have produced a spore which will exist as a TCM.

Effects on Consciousness by the Organization of Large Structures due to Black Holes

TL;DR: It is quite amazing how, when we zoom out far enough to see the structure of galaxies, they all have a gigantic black hole in their center. This Laniakea structure resembles that of a fungus.

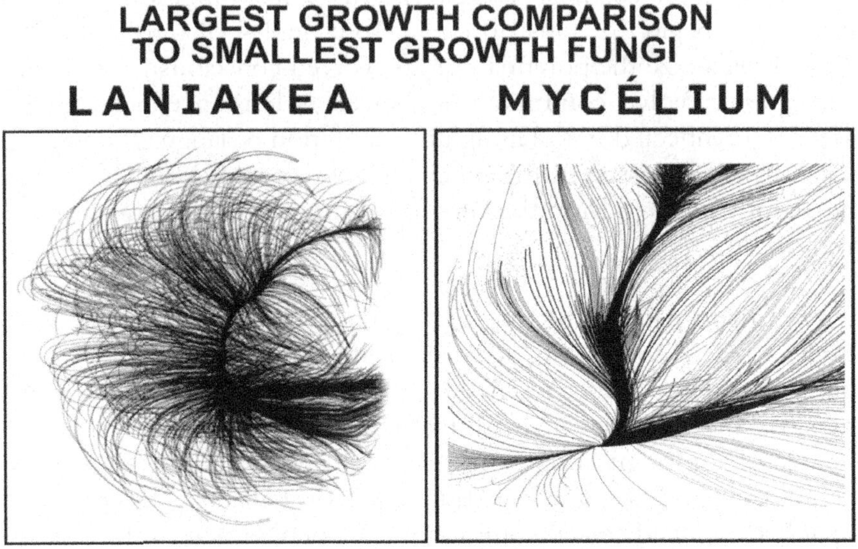

Figure H71P (Laniakea/Mycelium comparison)

This figure shows the similarities between Laniakea and Mycelium. The growth structures are the smallest and the largest and both are what a consciousness uses to grow the MCG.

Adaptability and Resilience with Extremophile Potential

ITM Fungus is always present and thrives underground where it performs many of the symbiotic tasks mentioned above; however, it is also ready to restart life above should a natural disaster (drought or radiation from a solar flare) or man-made disaster (nuclear annihilation) eliminate life above ground.

"A new study led by University of Illinois, Urbana-Champaign astronomy and physics professor Brian Fields explores the possibility that astronomical events were responsible for an extinction event 359 million years ago, at the boundary between the Devonian and Carboniferous periods."[82]

After all, and as has been shown before, fungi are likely the foundation of all complex terrestrial life and are the only reason life could grow back after such a cataclysmic event. Fungi are very resilient to extremes in radiation and other environmental conditions.

Scientists have known about this fungus, and similar extremophile organisms that can thrive on radiation, since at least 2007. The variety found in Chernobyl 'can decompose radioactive material such as the hot graphite in the remains of the Chernobyl reactor. The fungus grows toward the hottest and most radioactive places, like phototropism but for deadly toxins."[33]

Fungi are also capable of existing in very hot or very cold environments, which further reinforces the fact that they are prime candidates for helping ensure that life can continue to survive after catastrophic events. Some fungi can even survive in sub-zero temperatures.

"The majority of fungi are mesophiles, and grow at temperatures in the range of 5–35°C…Fungi exist, however, which grow very well outside the mesophilic temperature range; some are cold-tolerant (psychrotolerant) and are capable of growth near or below 0°C, while others are heat-tolerant (thermotolerant) and grow above 40°C.'[34]

Given that fungi are capable of surviving in such extremes, they are an ideal candidate for ensuring the survival of any biosphere, should disasters of a multitudinous variety occur on Earth. These organisms will act as a time-capsule, preserving some level of the evolution that had occurred up to that point. While other bacteria can also survive extreme conditions, they lack the networking and communicative capabilities of fungi that afford them the ability to balance ecosystems and sustain other organisms through their mycorrhizal networks.

IPT It has been claimed that fungi are one of the most, if not the most, important organisms for supporting life on Earth and are responsible for the evolution of multicellular life. To demonstrate the importance and vast array of capabilities of fungi, it is instructive to demonstrate how they are used today.

Unlike most plants, fungi are not capable of photosynthesis so they produce enzymes to absorb organic matter in their surroundings to generate energy for sustenance. This may be a mechanism in case the biosphere the fungus is growing on is cloud covered, and maybe even for millions of years, allowing it to inhabit first. They retain this ability, which animals also possess, as they were once merged with both plants and animals. Through this ability, fungi are extremely useful in breaking down various types of organic matter. Using fungi's ability to create enzymes to our advantage allows for the mass production of enzymes that are useful in a variety of ways.

"Fungal enzymes are compatible, efficient, and proper products for many uses in medicinal requests, industrial processing, bioremediation purposes, and agricultural applications. Fungal enzymes have appropriate stability to give manufactured products suitable shelf life, affordable cost, and approved demands. Fungal enzymes have been used from ancient times to today in many industries, including baking, brewing, cheese making, antibiotics production, and commodities manufacturing, such as linen and leather. Furthermore, they also are used in other fields such as paper production, detergent, the textile industry, and in drinks and food technology in products manufacturing ranging from tea and coffee to fruit juice and wine. Recently, fungi have been used for the production of more than 50% of the needed enzymes."[85]

More specifically, fungi are utilized for food items that are consumed daily all over the world. Yeast, a form of single-celled fungus, is used to make sure that bread rises. Some species of fungi are capable of producing rennin, which is used in the production of cheeses. The rennin produced by fungi alters the flavor profile. While cheese is also made with bacterial cultures most of the time, the products of fungi are still a part of the process. Some of the most well-known cheeses that make use of fungi are the blue cheeses Roquefort and Gorgonzola, which use Penicillium roqueforti. Yeast is also utilized in the process of making beer, which utilizes the Saccharomyces strain.

If the sound of Penicillium is familiar … it should be. This is the strain of fungi that the medicine penicillin was first discovered in. Shortly after its discovery, it was widely used as medicine to treat infections. Fungal cell walls and other constituent components of the fungi are polysaccharides. A polysaccharide is mainly a starch molecule, which is a number of sugar molecules bound together. These starches can be recognized by our immune system and are the key components for which an immune response is generated when an infection is present. Recombinant forms of these fungal polysaccharides are utilized to create vaccines. These are called subunit vaccines.

Another form of medicine that is becoming more mainstream is the ingestion of psychedelic varieties of fungi. The compound psilocin, contained within the Psilocybe Cubensis and other strains of mushrooms, is responsible for promoting the well-known psychedelic experience when ingesting these mushrooms. Medical science has recently become invested in treating depression, PTSD, and other forms of mental illness with psychedelic therapy. The results have so far been as good or better than the pharmacological interventions used prior, usually with little or no side effects. Research into other psychedelics has been given new life in the last decade after these compounds were listed as Schedule I drugs (the worst categorization possible and denoting the greatest amount of harm) in 1970 as part of the Controlled Substances Act. The truth is that most of these substances, like LSD and mushrooms, are far less dangerous than even the most seemingly benign painkillers we use daily. More treatments for a variety of mental illnesses continue to crop up, such as MDMA for couples counseling and ketamine therapy for PTSD.

"This is because there is a wide range of mycorrhizal fungal species that could change the strength of plant-plant interactions, and plant growth will vary. It is widely accepted that there are higher growth rates in plants inoculated with mycorrhizae than in control plants because of the increase in photosynthetic activities. Mycorrhizae also play an important role in the supply of essential nutrients to their associated plant; interestingly, fungicides or herbicides will not affect the growth of mycorrhizae."[86]

Here is a good example of the adaptability of fungi. This also shows evidence that this same substance is needed to spawn life on any biosphere that may differ in its bombardment and compositional makeup.

Since fungi are capable of producing a variety of enzymes, they can degrade and transform a variety of compounds. In nature, they will assist in the breakdown of dead plants and animals, however, that capability can be put to other uses. More recently, certain strains of fungi have been shown to be able to degrade plastic. Studies have shown that the marine fungus Zelarion meritimum possess the ability to break down polyethylene. Another fungus, Aspergillus tubingensis, possesses the capability to degrade polyester polyurethane. Fungi are also capable of absorbing nuclear fallout in order to cleanup nuclear waste. Specific strains of fungi have pigments within them that allow them to be increasingly resistant to radiation, as well as act as the absorbing compound within the fungi in order for them to clean up radioactive debris. There are some species of fungi without this pigment that are still resistant to radiation; however, the pigment enhances radioactive protection.

"Scientists believe this mechanism could be used to make biomimicking substances that both block radiation from penetrating and turn it into a renewable energy source."[87]

"Ekaterina Dadachova and her colleagues at the Albert Einstein College of Medicine in New York have discovered that some fungi can use a molecule called melanin, a pigment also found in human skin, to harvest the energy from radiation and use it for growth."[88]

Fungi are extremely capable life forms and have had a transformative effect on both plant and animal life. Many of the processes mentioned above would be much harder to implement, if not impossible, without the use of fungi. Such a broad range of capabilities make fungi a very powerful tool to utilize in colonizing and terraforming other planets into biospheres that can sustain human life. Experiments are already underway to determine if Mars can act as a viable host for bacterial, fungal, and plant life. Currently, it looks as if there is a positive result for growing one kind of bacteria that was pivotal in transforming the early Earth into something that resembles its state more today. With a bit more luck, fungi may also be found to be able to exist in Martian soil.

"The cyanobacteria grew well on the Mars substrate, and the nitrogen- and carbon dioxide-rich mixture at low pressure, albeit slower than in standard medium optimized for cyanobacteria. But this is still a major success: while standard medium would need to be imported from Earth, regolith is ubiquitous on Mars."[89]

BP Eventually, life on Earth as we know it will cease to exist. There are a number of disasters that could potentially wipe out all life on Earth. Some are more sure to happen than others. There is always the possibility of a large piece of space debris careening into Earth and wiping out life, as has happened to the dinosaurs. However, more assuredly, Earth will become uninhabitable due to the ever-increasing output of thermal radiation from the Sun as it ages (HCA). Due to this inevitability, humans will need to colonize both space and other planets. Elon Musk wants to send humans to Mars in order to start colonization. Given the fact that fungi are extremely capable and possess the ability to transform, degrade, and shield against a variety of harmful substances, it is no surprise that NASA is already looking into utilizing them to help in various construction efforts to assist in colonizing the Moon and Mars.

"Just like the astronauts, fungal mycelia is a lifeform that has to eat and breathe. That's where something called cyanobacteria comes in – a kind of bacterium that can use energy from the Sun to convert water and carbon dioxide into oxygen and fungus food. These pieces come together in an elegant habitat concept with a three-layered dome. The outer-most layer is made up of frozen water ice, perhaps tapped from the resources on the Moon or Mars. That water serves as a protection from radiation and trickles down to the second layer – the cyanobacteria. This layer can take that water and photosynthesize using the outside light that shines through the icy layer to produce oxygen for astronauts and nutrients for the final layer of mycelia. That last layer of mycelia is what organically grows into a sturdy home, first activated to grow in a contained environment and then baked to kill the lifeforms – providing structural integrity and ensuring no life contaminates Mars and any microbial life that's already there. Even if some mycelia somehow escaped, they will be genetically altered to be incapable of surviving outside the habitat."[40]

Given that NASA is already attempting to make use of these clever and resilient life forms, it can be inferred that their usefulness will only continue to grow and their applications become more diverse. In order to accelerate toward TES, being able to utilize nature to advance technology is obvious. Nature has assisted in many processes, as demonstrated prior, for up to thousands of years. With greater understanding, the fungus continues to better utilize its resources in order to ensure TES is achieved.

BBP Given the adaptability and resilience of fungi, and the fact that they show up early in the evolutionary timeline of Earth, it is essential for life to form elsewhere by other similar principles. No planetary body will house a biosphere that is immediately ready for the evolution of advanced multicellular life form the start. Transformation of the planet will need to occur in order to set up stable ecosystems and networks. Fungi represent the ideal organism to complete this task, and as such, it can be expected that planetary bodies around the universe will need to evolve organisms that mimic many, if not all of the same features of the fungi that have evolved here. All across the universe, fungi or fungi-like creatures are preparing their biospheres for the growth and evolution of increasingly conscious organisms that are ultimately capable of producing technology that can sustain the *fungus* indefinitely. Each of these fungi will be thrust into the cosmos after their host planet becomes uninhabitable. It is at this point that these fungi have become a TCM and are capable of interacting with other TCMs in space. This promotes MCG by allowing other cultures and technologies to interact and gain from one another, ultimately preparing the universe for the UGCH.

Interchangeability

A lab-created cell that is half human and half yeast. Researchers removed 450 genes critical for yeast's survival and replaced them with human versions of the same gene. They waited to see if the organisms would die but unexpectedly, almost half of the time new strains, which resulted, were capable of surviving and reproducing. While researchers have been able to swap single genes between yeast and humans before, this is the first time that hundreds of genes could be swapped on such a large scale. Researchers found that certain groups were stable over evolutionary time and suggest that there might be another 1000 "swappable" genes.

"It's a beautiful demonstration of the common heritage of all living things – to be able to take DNA from a human and replace the matching DNA in a yeast cell and have it successfully support the life of the cell," said Professor Edward Marcotte, from the University of Texas, Austin, in a statement.

Fungal-Human Symbiosis

TL;DR: Besides the fact that multicellular life started from the original fungus, it continues along with the conscious-growth process by re-integrating and having the innate ability to interchange as needed through the time evolving to reach TES. Surprisingly, the human body is made up of 54% fungus cultures and depends on such to survive. There is also the stoned ape theory that supports the development of the frontal lobe responsible for planning and imagination.

ITM It has already been discussed that fungi interact with plants through mycorrhizal networks in order to help sustain and enhance the nutrient uptake in the roots of plants. It has also been shown the plethora of ways in which fungi can help transform or degrade various compounds due to their ability to generate enzymes, similar to how animals digest their food after intake. However, fungi are also responsible for forming symbiotic relationships with animals as well. In the intestines of mammals resides a complex interplay of both fungi and bacteria which perform a variety of functions, which are not limited to digestive processes. Fungi in the human digestive tract are responsible for regulating immunity, both enhancing or reducing local inflammatory responses and regulating the balance between fungi and other bacterial organisms. This complex interplay, when in balance, promotes the health of the human organism.

"It is becoming increasingly clear that fungi are important components of the gut microbiota. Fungi residing in the human intestine, for example, elicit the induction of T helper 17 cells, which are central orchestrators of protective immune responses. Likewise, fungal members of the intestinal microbiota have been shown to influence the immunological responses of the mammalian host by dampening or promoting local inflammatory responses."[41]

It is interesting to note that these fungal and bacterial residents of our bodies exist in surprisingly large numbers. While a human has on average of 30 to 40 trillion cells that make up the entire organism, there are around 100 trillion or more cells of bacteria that reside in the human microbiome, which are the resident bacteria within the human gastrointestinal tract. Much smaller numbers of mycobiota (fungal organisms) are residents of the intestinal tract; however, they are significantly important to human health.

"Every individual's microbiome contains thousands of different species of microbes, of which 99.9% of the total number of microbial cells belong to only a few species. The less abundant (<0.1%), but more diverse, component of the microbiome has been termed the 'rare biosphere'...Humans have a fungal microbiome and it is clearly part of the rare biosphere of their entire microbiome."[42]

This demonstrates that even though the mycobiome (fungal microbiome) is significantly smaller in number than the microbiome (bacterial), its influence is significant. Though they may be small in number, their influence is very great in terms of the health of the organism. This excludes the consideration of the fungal communities that exist on the skin and other orifices, like the mouth. Thus, human-fungal interactions have evolved over long periods of time to promote a robust response to invaders as well as proper functioning of the human immune system.

IPT Given these facts, it is entirely likely that these gut microbes are essential for maintaining good health. That is to say that we have evolved from these fungi and other microbes but we also need them even to this day. The idea that we are somehow separate from them is a fallacy. Humans are of the very fungus that resides within them, which makes all life a single fungus.

"A stable gut microbiota results from a long co-evolution process aimed at supplying the host with numerous metabolic precursors, bioactive molecules, cofactors, and signaling molecules, neither primarily present in the diet nor produced by the host itself. Within human bodies Homo sapiens DNA is estimated to account for less than 10% of the total DNA because of the incredibly large numbers of microorganisms that reside in (and on) humans, primarily within the gut (Human Microbiome Project Consortium, 2012)."[43]

The development of the human brain, which is considered the pinnacle of evolution, relies on a very consistent supply of energy and other factors. These factors have been finely tuned to be produced by an ecology that mainly consists of fungi, bacteria, and other commensals. It is not a large leap of logic to make the realization that all of the evolution, in each species, has been so that the appropriate genes and biological capabilities have been cultivated in an effort to produce an ever-increasingly intelligent and conscious creature that fuels innovation. With expanded energy supplies for an increasingly complex brain, technological advancements are made possible by the interaction of these gut residents and their terrestrial vehicles. This creature is of the *fungus*, by the *fungus*, and for the *fungus*.

In recent years, studying the microbiome, and even more recently the mycobiome, has become a topic of great interest. Many diseases are now shown to be the result of a microbial or fungal byproduct (autoimmunity) or contain microbial/mycobial footprints in the gut, such as in the case of obesity.

"Microbial composition differs between people; for example, most studies have noted that the ratio of Firmicutes to Bacteroidetes is significantly higher in obese subjects... Furthermore, previous studies have shown correlations between an increased amount of Bacteroidetes in stools and weight loss, and between Firmicutes and development of obesity..."[44]

In modern times, changes in human lifestyles have had consequences on the number and type of bacteria and fungi that reside within the intestinal tract. These disturbances, which can be caused both by environmental factors, such as the usage of various pharmaceuticals, and dietary changes, such as the western diet's departure from natural foods and the adoption of highly processed foods, all have impacted the balance in the microbiome. Some of these changes are detrimental and cause a weakening of the gut itself, allowing for imbalance to lead to diseased states. Similarly, altered immunity can cause the body to change its inflammatory response, which changes the composition of the gut. This leads to improper gut function which can further drive disease.

"In recent years, several studies have highlighted the role of the microbiome in the pathogenesis of autoimmune diseases. Notably, in systemic lupus erythematosus an alteration of the intestinal flora (lower Firmicutes/Bacteroidetes ratio) has been described. Conversely, changes to the gut commensal and periodontal disease have been proposed as important factors in the pathogenesis of rheumatoid arthritis. At the same time, other autoimmune diseases (i.e. systemic sclerosis, Sjögren's syndrome and anti-phospholipid syndrome) also share modifications of the microbiome in the intestinal tract and oral flora."[45]

In some instances, pathogens can set off chains of events that cause the immune system to lose tolerance for things that were otherwise permitted prior. Thus, otherwise healthy people are brought to a state of sickness merely based on changes that occur within their microbiome.

"The human microbiome might be a major player in autoimmunity, as the loss of immune tolerance can be caused by microbial composition changes. Microorganisms can elicit the immune response against the host if the mechanisms of tolerance fail for several reasons. Recently, Rinaldi et al. found that autoantibodies directed against the cell wall mannan of the yeast Saccharomyce cerevisiae (phosphopeptidomannan), a ubiquitous commensal microorganism, were detected in several autoimmune diseases with different sensibilities (i.e. rheumatoid arthritis, systemic lupus erythematosus,

anti-phospholipid syndrome). More specifically, anti-S. cerevisiae antibodies (ASCAs) are a specific serological marker of Crohn's disease (CD) by appearing before CD onset in 32% of cases. In addition, S. cerevisiae is used as an adjuvant in vaccines, and this has led scientists to think of a hypothetical risk of developing abnormal immune activation that may be associated with an autoimmune/inflammatory syndrome induced by adjuvants (ASIA). Inflammatory bowel diseases (IBD) such as CD and ulcerative colitis (UC) represent an example of how the alteration of gut microbiome could induce disease. Notably, numerous studies have shown that both CD and UC are associated with a reduced complexity of the commensal microbiota and consistent shifts to a dysbiotic state."[46]

Given the fact that fungi and animals share a similar history, it would make sense that their DNA is compatible. Scientists have already carried out such an experiment with success! The words of the scientist that made this feat possible, Edward Marcotte, are quite profound.

"It's a beautiful demonstration of the common heritage of all living things - to be able to take DNA from a human and replace the matching DNA in a yeast cell and have it successfully support the life of the cell..."[47]

The ability for a cell to contain hybrid elements of both human and fungal DNA demonstrates that we are still genetically compatible with our fungal ancestors. Ultimately, what this suggests is that man must become more aware of its interconnectedness as one organism: the *fungus*. By doing so, many of the dysbiotic choices man has made over the last few hundred years will become the clear culprits for the increase in disease and dysfunction of the immune system. Fungi and bacteria are essential inhabitants that promote health and well-being when they are properly cared for and kept in check.

BP Given that the human immune system is highly sensitive to the balance of various microorganisms in the microbiota, it is almost a surety that advancements in medicine and treatment technologies will hinge upon understanding the importance and power of various species of fungi and bacteria. The fact that immune antibodies to certain yeasts are produced by the human immune system is indicative of their role in health conditions. Perhaps, with the use of gene editing technologies like CRISPR, it will be possible to custom-craft specific and unique communities of microorganisms that are safe for specific individuals.

Essentially, each patient could have a custom-tailored microbial community that helps balance various inflammatory responses and protects them from foreign invasive microbes they may encounter from bad food or other infectious events. Humanity is at its early stages in terms of understanding the importance of these organisms that have lived for millions and billions of years. However, as we progress and are taken back to our roots, we come full circle and expand our consciousness. Once again, we see that the goal of the *fungus* is to become more aware and then generate technologies that can improve the human condition. This is the process by which TES will be reached and the survival of the *fungus* is ensured. In this particular instance, it is the recognition that all organisms on Earth form one greater organism through the constant balance of dysbiosis and symbiosis. With greater understanding, greater levels of symbiosis will be achieved.

BBP Given that fungi exist in a tenuous balance of symbiosis and dysbiosis within humans, it is obvious that their power over us cannot be ignored. Civilizations with conscious awareness and technologies that far surpass those of modern humans will have both mastered its effects on their health, as well as the aforementioned properties of resilience and adaptability. These civilizations, as well as future humans, will leverage the abilities of fungi to their limits. They will organically grow various fungi to generate structures that are resilient to radiation, use these creatures to easily decompose unwanted materials, and even have gene-edited various species of fungi to produce medicines, materials, and food. The importance of fungi cannot be overstated and this will continue to be true for humans that have achieved TES and are existing as TCMs.

The spore of the *fungus* will continue out into the cosmos, taking each and every bit of knowledge it has learned, both technologically and organically in order to bring about MCG. With further interactions with other civilizations of *fungus* that have achieved ESS, technologies can be shared to bring about new understanding. This is the process by which the *fungus* reaches its penultimate state and serves as preparation for the universe's end goal: the UGCH.

EOS

- Fungus is responsible for the digestive process.
- Fungus is directly related to brain clarity.
- Fungi DNA is closely related to human DNA.
- Humans are directly related to ancient fungus.

The Right Tool for the Job to Grow Consciousness

TL;DR: If the goal was to develop a consciousness of very high value, you could only do it by tricking it through the growth process in real time. By obscuring the evolutionary evidence from its conscious state, it is allowed to move forward in time unconsciously, ignoring the end goal of extraterrestrial existence until it is very obvious. To evolve the span of time needed to reach TES and hide conscious-growth from its true purpose and origin, growing it inside fungal matter is the best physical method.

ITM Early forms of algae were responsible for changing the atmosphere of the Earth and formed symbiotic relationships with fungi. They were responsible for transforming a lot of the gasses and minerals into oxygen, which helped create the atmosphere that would give energy more efficiently in the form of breathable air to animal life. Fungi have helped early plants gain access to more nutrients as well as communicate with them in order to alert them of invading and hostile organisms.

"And, according to a team of researchers from the University of Leeds, those early plants weren't developed enough to do it on their own. They had not yet evolved roots or vascular systems like plants have today. Instead - and not dissimilar from symbiotic relationships between plants and fungi today - fungi in the soil transferred phosphorus from the rocks to the plants, which in turn powered plant photosynthesis."[48]

Fungi have helped mankind tackle a variety of culinary and medicinal challenges as they have evolved. From breads to brews to blue cheeses, fungi have utilized their enzymatic abilities to transform various compounds or generate others that have been extremely useful to man. No other organism in the history of the Earth possesses the vast array of capabilities to interconnect, support, communicate, and transform materials the way fungi do. Accordingly fungi have found their way into myriad ecosystems in order to enhance and support them. In most cases, this was not a competitive encroachment but a necessary symbiosis that allowed life to thrive. For this reason, all subsequent life to fungi is essentially still a fungus. The genetic changes that took place in the organisms that branched away from fungus may make these organisms seem unique and distinct; however, they all interact with fungi in very profound ways. In essence, fungi are interlinked with everything that exists today. Everything is still part of that *fungus*, and thus, all life can be considered a large and networked *fungus*.

IPT Man continues to utilize the vast array of capabilities for which fungi are well adapted for their culinary endeavors. New medicines are created utilizing fungi by looking at variations of their polysaccharides and loading them into vaccines. Fungi are being considered for material construction on other planets as well as shielding buildings, suits, and spaceships from the harsh radiation of interstellar space. It makes sense that the *fungus* is trying to prepare us to leave our planet when we see the many ways we try to utilize literal fungi to advance our material sciences enough to withstand the harsh environments of outer space and other planets.

"The myco-architecture project out of NASA's Ames Research Center in California's Silicon Valley is prototyping technologies that could "grow" habitats on the Moon, Mars and beyond out of life – specifically, fungi and the unseen underground threads that make up the main part of the fungus, known as mycelia."[49]

BP From the appearance of the first fungi to the present day, fungi have been transforming and enhancing the atmosphere and networking with all life on the planet. This vast network of organisms, the fungus, is perfectly adaptable and capable of self-correcting and finding new homeostasis when the whole ecosystem is perturbed. It is this aspect that makes the fungus the perfect tool for the growth of consciousness. With its ability to network, it stores information and adapts to change. With its resilience and adaptability, it can survive in, and make use of, materials in many different environments. With its ability to transform and evolve, it finds newer and more efficient ways to evolve. By the time the Earth is no longer habitable, this *fungus* will have priced in the dangers far in advance by utilizing the knowledge of all the networked organisms, both long gone and still living, in order to move toward greater capability and a greater capacity to solve problems. Ultimately, this leads to the production of a spore within that *fungus*, which is the BCT. This spore will be responsible for bringing the *fungus* to safety when the Earth is no longer habitable. By viewing this evolutionary process from beginning to end, it is obvious that fungus was the right tool for this job.

BBP Anywhere life should spring up in the cosmos, a fungal system will need to be generated very similar to what has been generated on Earth. This fungus will be responsible for transforming the ecosystem of the biosphere it inhabits, making it ready for an interconnected web of evermore efficient organisms that utilize the materials found in both the ground and atmosphere to achieve greater levels of consciousness, awareness, and intelligence. The *fungus* will push that life forward, hopefully achieving TES and producing its own spore. These spore TCMs from each of these extraterrestrial biospheres will go on to populate and network within the cosmos. Each *fungus* will then have the possibility of interacting with others, expanding culture, knowledge, and understanding. All of this will promote MCG as the universe heads toward the UGCH.

Technology Grows Like a Fungus

TL;DR: Technology is the direct result of a fully evolved *fungus*. Technology itself evolves much like the behavior of fungus. In that every simple or complex idea is transferred from one human brain to another, in the billions over the years or centuries. The spreading of the logic of innovation is information fungi. Einstein may have been an individual spike in the collective with an ability to dive deeper than most, but he is also the direct result of a collective conscious-growth before him, launching him to great depth of thoughts. The collective consciousness spawns such individuals that continuously accelerate its conscious-growth. It is a recurring process. Spikes produce conscious acceleration and a stagnant consciousness gives way to spikes.

ITM Throughout the evolution of life on Earth, even the earliest single-celled creatures evolved different methodologies for defending themselves, gathering nutrients more efficiently, and developing various evasive (locomotive) techniques. Each of these things could be considered life's first attempts at generating some form of technology. While they are organic in nature, some of the technology we employ today is also organic in nature and not some machination. More importantly, these early advancements in life also grew outward and upward. Sometimes these traits were passed down to future lineages (vertical gene transfer) if they were beneficial to the adaptation of the organism. Sometimes these traits were even transferred to entirely different species (horizontal gene transfer). This progression can be likened to that of a network or branching structure, which often is used in looking at genetic lineages or even your family tree. In that sense, this early technology employed by these organisms is obviously fungal in nature. In addition, the same can be said about more complex multicellular organisms. The elongation of beaks in some birds in order to be able to better extract nectar from various flowers, or the change in color of fur in animals, are other examples of traits that act as a form of technology that allows these organisms to be better-suited to their environment. This process continued up until the evolution of the first BCTs. It is at this point that the technologies we more commonly attribute to that word began to evolve. At some point, animals began utilizing their surroundings without making changes to their DNA. This was the beginning of the kind of technology we are more familiar with today. For example, the first animals to burrow into the ground in order to stay warm during extremely cold periods were using the Earth as a tool in order to survive. This is an example of organisms making use of outside materials to their advantage. There are other examples as well; however, it wasn't until animals could store the things they learned in some form of language structure that was somewhat permanent that technology began to grow exponentially.

At some point, early humans began to store information not just through spoken word but through written language. Scrawled on the walls of caves and other structures, and then later on papyrus, information was able to be recorded and passed down and collected from generation to generation. This kickstarted the exponential growth of technology. Like the networking and branching structure of the mycorrhizal networks that fungi utilize, technology began to sprawl outward from some starting point and progressively improve. Ideas could then be combined in order to spawn innovations that made early man more successful and adaptable to his environment.

IPT Currently, man has advanced forms of information storage (e.g. immediately after the internet, extremely large hard drives and solid state drives, and even DNA-based storage) that it utilizes in order to transfer the wealth of knowledge it has gained from one individual to the next. Storing information this way has allowed humans to evolve from sending messages over the oceans by boat or across the land by horse to being able to send messages instantly with the click of a button on an electronic device. People are able to learn about each other and about things they wish to pursue within education instantaneously. Humanity can now offer individuals entire degree programs online. Elon Musk currently wishes to expand the ability of humans to access the internet, even in otherwise prohibitive areas where infrastructure is limited. Starlink, a satellite-based internet that Elon Musk has been building and testing, is the latest form of conscious fungal communication in the sense that it aims to connect all humans around the globe to instantaneous information in real time. Starlink is an example of a technological network for communication that possesses fungal qualities. Similar to mycelia, which spreads nutrients and communicates with other plants, this network will spread information through its various networks and allow others to communicate over vast distances. This allows knowledge and technology to continue to grow exponentially.

BP From the harnessing of fire's power to the development of rockets and rocket fuel, all innovations work together. When viewed from a broad perspective, each technological advancement appears to have branched off from prior technologies, the way a fungal network would branch out as it grows in order to find new food sources. If technology were to be viewed in a time-lapse over decades at a time, as what can be observed by watching a cityscape evolve from pre-industrial times to our current modern infrastructure, it would seem to grow like an organism exhibiting uncontrolled and rapid growth. It's a slow process but after the advent of the computer and the internet, technology begins to accelerate rapidly. Life on a biosphere only has so many years to advance during this period of technological advancement, as much time has already passed since the formation of the planet; leaving much less time on the clock in which the *fungus* can achieve TES. Thus, the last legs of technological advancement occur in a rapid and hurried manner in order to ensure the survival of the *fungus*. By this point, the tendrils of technology will have explored the majority of areas required to generate the specific innovations that assure TES. Each progressive step forward trains the *fungus* in adapting better to its own environment as also to other environments. When TES have finally been achieved, the *fungus* will have reached the spore stage and mankind will leave its biosphere as a TCM.

BBP All throughout the cosmos, various fungal ecosystems will have been thrust away from their terrestrial birthplaces and into the vast expanses of the universe. These separate networks of *fungus* may co-integrate like the mycorrhizal networks that exchange nutrients and communicate with the root networks of various plants. This co-integration would allow for the sharing of vast amounts of information, each an entire civilization's history in size and breadth. It is in this manner that these TCMs may continue toward MCG for the UGCH.

EOS

- Birds, beetles, ants, and primates use sticks as technology.
- Fire to moon landing.
- All technology connected from the advent of flint forward.
- Animal evolution to bipedalism freeing the thumb.
- All and every innovation/invention is for the support of.
- Technology being the only thing to save life from a biosphere ending.

Fungus Theory

TL;DR: It is more a fact than theory that all biology exists as one organic entity with the purpose of succeeding post-terrestrial. Aside from the difference in composition of other biospheres out in the universe, all intelligent life should spawn and grow and succeed from a *fungus*. This evidence suggests that all organic life on a biosphere is one self-feeding *fungus* (powered only by the host star), in the sense that making an animal feed off of plants or other animals, is a form of the *fungus* feeding itself. In other words, everything relies on its own adaptive growth in order to feed itself. This can be looked at as a unique form of cannibalistic growth toward intelligence.

It has been demonstrated that fungi have been integral to the evolution of the ecosystem on Earth and have made it habitable for increasingly complex and evermore conscious organisms. Fungi continue to integrate with all forms of life in a variety of ways, promoting interconnectivity which expands the conscious awareness of all life forms as a byproduct. As such, all life is some variation of fungus and should be considered as such. If every subsequent life form to fungus is connected with fungus, sharing much preserved genetic ancestry with fungus, then it would not be any stretch of logic to state that it is all one *fungus*. This *fungus* aims to prepare a spore (a bipedal organism) that is capable of achieving the technological requirements necessary for evacuating its birthplace. Once TES has been achieved, the *fungus* ejects this spore into the cosmos in order to survive and thrive. What happens next is anybody's guess; however, that form of consciousness will continue to adapt and maximize its potential under new conditions and in its new surroundings. In essence, *fungus* promotes MCG at all times and is optimally suited to do so. Can be looked at as, from the mother cell to TCM is one long, single and fungal DNA strand.

BIOLOGICAL/*FUNGAL* EVOLUTION OF CONSCIOUS GROWTH

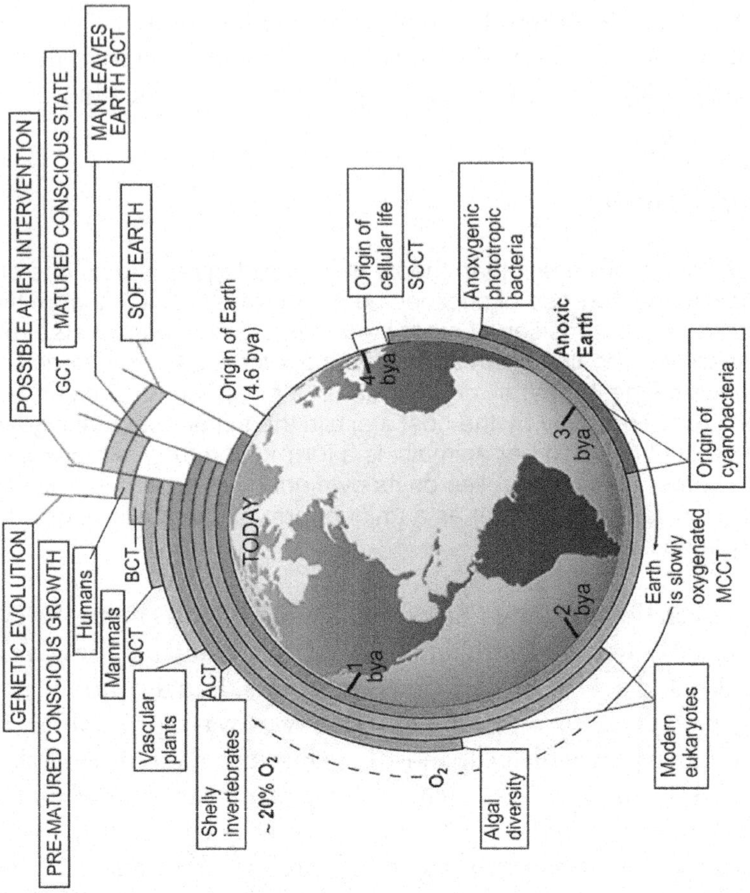

FIGURE 1.6A

Figure 1.6A (Biological/Fungal Evolution of Conscious Growth.)

This figure shows the entire biological growth process on Earth from beginning to single cell, multi cellular, and evacuation. All the many different cellular evolutionary processes that must occur biologically for that biology to become intelligent enough to escape its doomed birth planet.

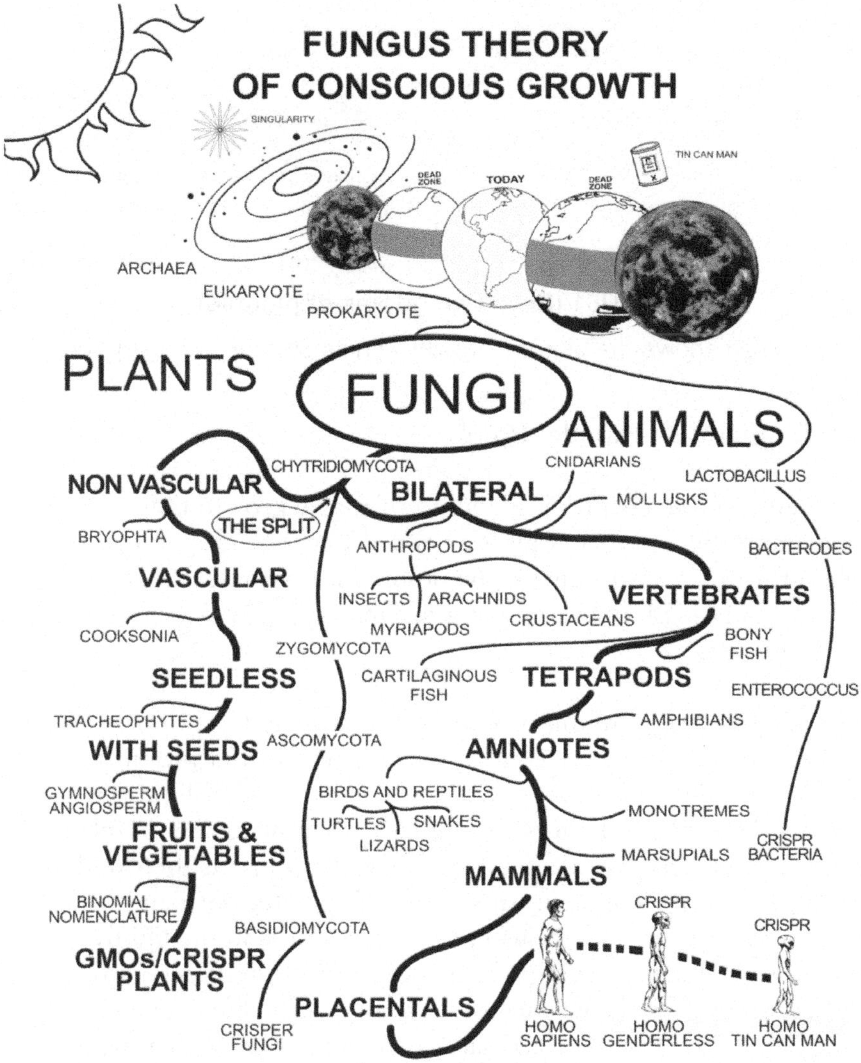

FIGURE F21

Figure F21 (Fungus Theory)

This figure shows the bacterial building process that is the foundation for the development of Fungi. This Fungal process is how a consciousness evolves by the end of its terrestrial beginnings. The forensic evolutionary evidence of how intelligence evolved on Earth. A self-feeding, solar evolutionary process that limits the awareness of what may become conscious and have the ability to trace back to its true origin.

Linnaeus 1735[32]	Haeckel 1866[33]	Chatton 1925[23]	Copeland 1938[7]	Whittaker 1969[34]	Woese et al. 1990[35]	Cavalier-Smith 1998[30]	Cavalier-Smith 2015[36]
2 kingdoms	3 kingdoms	2 empires	4 kingdoms	5 kingdoms	3 domains	2 empires, 6 kingdoms	2 empires, 7 kingdoms
(not treated)	Protista	Prokaryota	Monera	Monera	Bacteria	Bacteria	Bacteria
					Archaea		Archaea
			Protoctista	Protista	Eucarya	Protozoa	Protozoa
						Chromista	Chromista
Vegetabilia	Plantae	Eukaryota	Plantae	Plantae		Plantae	Plantae
				Fungi		Fungi	Fungi
Animalia	Animalia		Animalia	Animalia		Animalia	Animalia

Figure U81 (Seven kingdoms of plantae)

This graph shows the plantae growth with its corresponding epoch.

Early Fungus Was Bipedal Prep

The material in this chapter is evidence of fungal growth that demonstrates the fact that plant life needed to concentrate nutrients by evolving for millions of years to the point where there existed fruits and vegetation to nourish the more complex and energy-hungry brains of bipedal organisms.

The underground communication network of fungi is a prime example of why the interdependence of various fungi and plants within forests are needed for the development and ignition of terrestrial life forms. In the evolution of plant life, it is vegetables that show up last. While it may seem like a coincidence that vegetables show up just before the development of bipedal organisms with their greater nutrient requirements, it is not. Fungi also play a critical role in the tightly-restricted food chain by living off the decomposition of dead and dying plant matter. This evidence suggests that all organic life is one self-feeding *fungus*, in the sense that making an animal feed off of plants or other animals, is a form of the *fungus* feeding itself. In other words, everything relies on its own growth in order to feed itself and can be looked at as a unique form of cannibalism. It's evident that the timing of the arrival of both the BCT and the higher nutrient density afforded by vegetables being so close in proximity to one another emulates the fungal processes described previously (e.g. immediately after mycorrhizal networks) but on a grand scale. That observation demonstrates the tenacity of all plant, animal, and technological growth behavior as the fungus moves forward to its intended evacuation goal, or TES, without being obvious to consciousness as it grows. Thus, the classification of plant, animal and fungus as separate kingdoms is irrelevant. These are representations of one large continuous fungal-like organism. This construction process is essential in hiding true origin from conscious-growth when it eventually evolves to the point of sentience and should be similar on any biosphere throughout the universe. The behavior of fungus in general and the fact that it is everywhere and on everything supports the idea that it is the foundation of all life and all life may be one large long-lived fungus – which summarizes Fungus Theory via this book.

3. EVOLUTION

The Evolution Of The Bipedal Computer Transporter (BCT)

Introduction

This chapter takes a look at evolution from the viewpoint of the formation and improvement of a specific structure: the brain. It aims to illuminate the processes by which early life progresses in order to generate a bipedal organism-the bipedal computer transporter (BCT) that is capable of developing the necessary technologies to ensure the survival of the *fungus*. In order to help the reader understand how the *fungus* arrives at the BCT, an account of the various steps that the *fungus* necessarily iterates through, in order to produce more efficient designs, as well as prepare sources of food that become more nutrient-dense, is detailed. The reader is also made aware of the importance of the anatomy of the BCT when compared to the other organisms that came before it, and how it is pivotal to the development of any meaningful technologies in order to ensure the end of terrestrial survival of the *fungus*.

The general progression that will be covered in further detail is that the SCCTs (like bacteria, yeast, and algae) transformed the environment, seeding the biosphere with molecules like oxygen that were to serve as an energy source for the MCCTs that would come thereafter. The continual evolution of the MCCTs required greater capacity for the storage of information, which resulted in the formation of the first brain-like structures. ACTs evolved after the MCCTs and represented a leap in processing capability as they utilized the structures that had evolved in SCCTs and MCCTs in order to create a specialized set of organs for processing and sensing. This new organ, a specialized computational organ (a brain), along with the explosion of aquatic life, allowed for an extreme expansion of variation in life. Eventually, these ACCTs would be pushed toward land to escape the extreme competition of aquatic life. Those that evolved and survived on land became the first QCTs. With time and further evolution, the QCTs gave rise to the BCTs and not long thereafter the BCTs began developing technologies with their sophisticated brains. Each one of these brain transporter organisms have shown to be a necessary growth pattern for this *fungus*' development of a brain capable of ESS.

"Every living thing is, from the cosmic perspective, incredibly lucky simply to be alive. Most, 90 percent and more, of all the organisms that have ever lived have died without viable offspring, but not a single one of your ancestors, going back to the dawn of life on Earth, suffered that normal misfortune. You spring from an unbroken line of winners going back millions of generations, and those winners were, in every generation, the luckiest of the lucky, one out of a thousand or even a million. So however unlucky you may be on some occasion today, your presence on the planet testifies to the role luck has played in your past."
– Daniel C. Dennett, Freedom Evolves

All Life Is for the Development of a Highly Conscious Brain

TL;DR: Like the behavior of slime mold or fungus, the *fungus* grows a brain capable of successfully leaving Earth as an eternal self-sufficient animal (ESS) through a gradual terrestrial evolutionary process. This assures that that brain is one that has had a chance to be each and every animal life form as a growth catapult to its intended capabilities. By the end of terrestrial supported life on Earth, the *fungus* (all organic life) will have constructed this with the growth of that brain's conscious abilities to make reality of its environment's limitations. This process also doubles as a method of the cognitive inability to decipher that consciousness's True Origin (TO).

The main function of evolution is to develop the capabilities of a highly-evolved brain, specifically one that is capable of accepting more sensory input and responds more to its environment through the interactions of the environment with the body that houses and is connected to the said brain. No matter the size or presentation, it is an organ that is capable of accepting the sensory input of the surrounding environment from a variety of sensory organs that are connected to it. These sensory organs will be excited by the stimuli of the surrounding environment without affecting the underlying signal or performing any kind of processing on these signals. They simply report what they detect. All life forms have some form of system that functions in this way, whether that system utilizes a complex brain or not. Even single-celled life forms seem to possess memory when exposed to various environmental stressors. This memory is likely stored in the nucleus. From the very first single cell dividing on Earth, all the way up to the evolution of bipedal organisms, the entire focus of Earth's organic evolution drives toward the development of a brain-a highly sophisticated one at that. From single-celled organisms demonstrating functional memory to multicellular organisms developing complex structures to house their consciousness, these "nuclei" of information and memory must find a way to eventually become upright (bipedal) in order to optimize their size, function, and as a result, their sensory capabilities. The input of information is limited by the physical configuration by which it is transported. The increase in sensory capability of these organisms, through changes to its physicality, provides the raw data with which a brain can evolve. The more information the brain can receive, the more it can do with it.

This is an important aspect to the evolution of the *fungus* because a variety of different experiments are being run simultaneously. Multiple strains of the earliest bacteria existed on the Earth, each accepting the inputs of their environment in slightly unique ways. Continue this evolutionary process up to modern animals, where some perceive sounds that humans cannot hear or be attuned to frequencies of light outside the visible spectrum, and you see the scope of the possibility space that the *fungus* is exploring. Each of these differing configurations of what signals may be sent to the brain promotes alterations that the brain will undergo in order to transmit and receive these signals more efficiently or with greater sensitivity. Thus, we have an entirely physical mechanism that describes one of the reasons behind the evolution of the brain.

EOS

- Everything starting with one original cell
- That one cell evolving into billions of other cells
- Ending up as one cell again, having evolved from every previous cell

Evolution of, and Mechanisms for, Developing Increasing Intelligence

TL;DR: It takes a wide diversity of animals to keep the growth of intelligence to advance. The three main types are primary, secondary, and tertiary consumers. One emerging in the food chain at a time and in that order shows how the fungus advances stage by stage. This system, the growth of becoming more and more intelligent to escape your predator, is the driver to eventually becoming bipedal (conscious) and off the planet as a spore.

While the life generated by the universe has the inherent goal of creating organisms with increasing levels of consciousness, merely being aware is insufficient for making a robust and resilient organism. While consciousness serves as the ground level or foundational layer for any organism within the same species, it is its ability to employ a variety of tasks that relate to processing that sensory input that comes from consciousness, as well as the efficiency with which that input is processed, that delineates differing levels of intelligence. Thus, the path to a BCT is one of increasing levels of consciousness through sensory input, processing of those inputs through a centralized brain, and the increasing efficiency with which all this processing is done. Without each of these aspects of evolution, the task of generating an intelligence that is capable of solving the problems required to reach TES (Technological Evacuation Standards) would be insurmountable.

Most people today would not think that slime molds are intelligent; however, they do possess the ability to take the information from the environment and adapt to it. This behavior seen in slime molds demonstrates that the beginnings of intelligence occurred very early on in our evolutionary history. This example demonstrates that some form of computational processing is being done which informs the entire collective organism to react in a way to improve its current state of affairs. While slime molds are single-celled creatures, their ability to communicate with other members of the collective demonstrates a form of intelligence that allows for interesting behaviors. Refer back to Chapter 2, where slime molds were able to solve the Tokyo subway transit problem more efficiently than humans, when presented with the grains of oat scattered in the same pattern as the major hubs/cities in Tokyo.

This intercommunication is a crucial aspect of intelligence, but it has its limitations. Each cell in the slime mold is essentially a copy of all the others. Thus, there is no specialization for different types of tasks or processing. As evolution continued to occur on Earth, multicellular life arose and this allowed for individual cells to perform specific tasks. Instead of every cell having the same job, some cells were specialized for receiving information from the outside world whereas other cells were utilized for locomotion or defense. This specialization allowed for processing to occur within more localized regions of the organism. While signals are sent from all around the organism, they are processed in specialized cells. This is the point at which the fungus had iterated toward a proto-brain. With further evolution, these cellular structures for processing became more advanced and more efficient. The eventual result of this continued evolution of specialized cellular proto-brains resulted in the first animals to house primitive brains. These organs are the main control centers for these animals and require a large amount of energy to continually maintain function of the organism. This specialized organ is now responsible for the formation of memory, information processing, management of biorhythms, and responding to a variety of stimuli.

For any given species of animal, a variety of appendages and sensory organs provide signals for the brain to receive. The brain must then take these signals and process them in a way that doesn't overwhelm that organism. For example, if a human had access to every bit of data streaming through its brain that related to monitoring heart rate, digesting food, reinforcing the immune system, cleaning the blood, and a whole slew of other signals, humans would be quickly overwhelmed. It is the process of segmenting, compressing, and compartmentalizing this constant barrage of sensory input that drives at the heart of the mechanisms behind intelligence.

In order to be more intelligent, an organism must not only be receiving signals from the environment, it must process them in order to find important relationships between them. It would be useless for an organism to be receiving signals that the temperature of its environment was rising if not to integrate those signals into a higher level of cognition. Likely, this information would inform the organism that it must make changes to its current situation to maintain survival. This basic example illustrates that further processing is required in order for something meaningful to come from sensory input.

With greater levels of processing, and the interconnectivity and cross-signaling between various networks found in the evolving brain, more nuanced and unique "features" of consciousness are generated. Being able to relate how one "feature" relates to another, and produce actions which promote seemingly desirable outcomes for the organism are the underpinnings of the evolution of intelligence. In a sense, when the brain is now occupied with not only ingesting the raw sensory input but also processing these inputs in conjunction with one another, a new modality of behavior in that organism begins to become apparent. Behaviors like learning, memory, and adaptation all begin to become much more complex than the behaviors of single-celled organisms, early multicellular organisms, and early animal life. The hallmarks of this type of intelligence can be seen in the explosion of technology, written language, manipulation of materials, and artistic expression found in ancient cave paintings and architecture. Each of these is evidence that the brain has evolved to such an extent that it is no longer concerned merely with survival, but has focused on ways to enhance the surrounding environment beyond simple, organic, chemical interactions or employing basic burrowing techniques. These achievements are technical in that they have meaning, require planning, and demonstrate understanding of advanced concepts like geometry and physics. In order for an organism to engage in these behaviors, it obviously takes more than processing and responding to external stimuli. As mentioned before, multiple features are generated from these external inputs (e.g. abstract thinking, internal reflection, hypothetical postulation, and recording of information for future generations) that allow the organism to do more than react to its environment.

As humans continue to evolve, more modalities of behavior will continue to come to light. These new behaviors will allow for different levels of interaction between humans that are not common, or may not even exist today. This may occur due to the advent of virtual environments, or it may occur because humans decide to upload their consciousness entirely into organo-computer networks. The only thing that has been consistently seen throughout the evolution of the fungus is that increasing the consciousness of the organisms (the raw sensory inputs made available to processing centers) results in increasing intelligence through the novel interactions that these organisms will have with their environment. Over time, these interactions stimulate these processing centers to evolve toward becoming more efficient and allowing a greater contribution toward the consciousness of the fungus as a whole. Each time the fungus becomes more conscious, it also becomes more intelligent by spawning organisms that utilize this enhanced consciousness to exhibit new forms of processing and behavior. This process is incremental and spreads from one organism to another in a fungal manner, where some branches may end and require much less iteration beyond a certain point, and other branches continue to expand outwards much further due to the perceived benefit in that direction. If one were to observe all life as having sprawled outwards from one point in a top-down view, the different iterations and evolutions of different creatures, some having gone extinct early on while others endured for much longer, one would perceive something that looks much like a slime mold traveling toward different sources of energy in order to survive. In the view of Graham Cooks, chemistry professor at Purdue, it is a dramatic discovery.

"This is essentially the chemistry behind the origin of life," he added. *"This is the first demonstration that primordial molecules, simple amino acids, spontaneously form peptides, the building blocks of life, in droplets of pure water."*[50]

SCCT – The Terraformers

TL;DR: It would be obvious and detectable if life were to exist or evolve in any other method, giving way to the discovery of where that life and its true purpose began, canceling out the process of conscious-growth. Starting with a single cell on any biosphere is the perfect way to hide the detection of true origin with grand changes to that first cell over time and mutation. It also needs to adapt to the specific composition of that biosphere which may differ from biosphere to biosphere throughout the universe.

Bacteria, some forms of fungi, and other early life forms started out as single-celled organisms and were responsible for the vast majority of transformations to the atmosphere that would promote the further evolution of life. Specifically, cyanobacteria and algae were responsible for using photosynthesis to generate vast quantities of oxygen. These organisms utilize carbon dioxide and sunlight to produce this oxygen, thereby reducing (or fixing) carbon dioxide. By increasing oxygen levels and reducing carbon dioxide levels, these early organisms paved the way for the more complex forms of life that would come afterwards.

The importance of these early life forms cannot be overstated. Without the initial reduction in carbon dioxide, the greenhouse process on Earth would have run rampant. By generating large amounts of oxygen, these organisms allowed for the formation of an ozone layer. The ozone layer is purely made up of oxygen in the form of O3. This particular molecule happens to scatter ultraviolet light. Without this layer, life on Earth as we know it would not be possible as the ionizing radiation from the sun would be too damaging for land mammals and the majority of plants that were to follow.

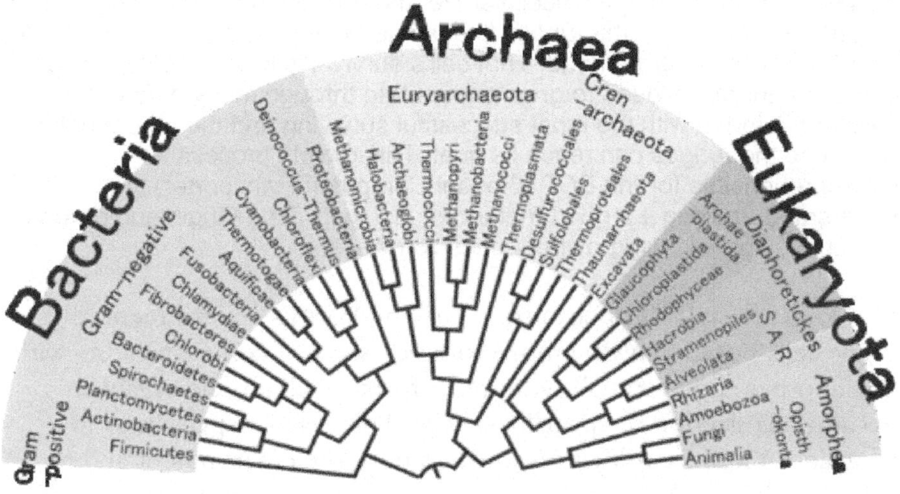

Figure 99 (Phylogenetic Tree)

This Phylogenetic Tree of Life is the best example of how all life starts from amino acids that become the building blocks for the starting of a *fungus*. This bacterial development is necessary and specific to form the correct fungal composition on Earth's particular biosphere. Once this foundation is established, the fungus will take off like a bandit starting the many fingers of life forms (feeding on one another) needed to grow a consciousness capable of technological evacuation of the very biosphere that cultivated it.

Fungus' Primary Goal

The *fungus* knows it has only so many years to get off the planet once it starts. The SCCT stage of the fungus is essential for the foundation of all organic matter and animals that will evolve after. It is responsible for consolidating and transforming all the compounds that exist on the biosphere into the types required for the next stage of organisms that the *fungus* will go on to produce. This early stage of life lays the foundation for the perpetual growth of consciousness, ultimately driving the entire fungus toward MCG.

EOS

- Single cell animals communicate together in order to find food.
- Single cells cluster together to make it appear larger for protection.
- The larger cluster also reaches out to find food.
- A conscious effort to survive.

MCCT - The Transformers

TL;DR: After becoming multicellular life, the development of a memory cell within that cell. This is the *fungus*' first stage in developing a brain that will later form as a separate head. That cell's survival is key to the conscious development of the need to grow and mutate through many different cellular life forms with the most successful surviving and the others going extinct so the *fungus* can reach its goal. This growth process is incomprehensible for the BCT (Bipedal Computer Transporter) to understand, making it a good method of hiding its True Origin and purpose and continuing its own growth of consciousness.

Not long after SCCTs began transforming Earth's environment, these single cells began evolving more complex ways of communicating with one another and eventually adopted a form in which one organism became a collection of specialized cells. This cellular specialization allowed for a more robust response to both the environment and to other organisms. Cellular specialization is what begins the process of being able to generate uniquely evolved structures that operate for some subset of functions in the greater organism. This is the point at which early organs began to evolve. These MCCTs were more efficient in finding resources and processing information, because of their increasing complexity and the variety of tasks that could be performed, due to the benefits of cellular specialization and subsequent organ development. By allowing for specializations of individual cells in these organisms, a vast array of new evolutionary options became available to MCCTs.

This form also has the added benefit that a potential threat to a single-celled organism may no longer be a threat to a large collection of cells. This is the case due to the fact that the size of the organism is increased, making it harder for predators to mount attacks. As a result, an increase in the robustness of the organism occurs as some cells may be damaged or destroyed without killing the organism. By protecting themselves from these threats, an evolutionary advantage was afforded to these MCCTs that allowed them to continue to evolve and promote the next form of life in the evolution of the fungus.

Thus with the increased size of the organisms, the energy requirements for survival also increased. It is at this stage of the fungus that increased locomotor capabilities also evolved to allow these organisms to scavenge for resources in order to satisfy their larger energy requirements.

Fungus' Primary Goal

By this point, the *fungus* is adapting to various environments more easily due to the advantages afforded by cellular specialization. The fungus moves in this direction in order to increase its ability to gather nutrients and protect itself from the dangers of the environment. By being able to evolve appendages that are specialized for movement, gathering of nutrients, and surrounding/protecting vital structures, the fungus has iterated one step further from the SCCT into a new form that houses a plethora of options for adaptation and anatomical configuration. This is a critical stage in the evolution of life for the fungus as the number of options available to it for experimenting and iterating through has grown exponentially. This is a profound leap in pushing toward MCG as consciousness can now be housed in more than one cell.

EOS

- First eye development
- Development of complex organs
- Better means of mobility
- Better defense capability

ACT - The Skeletors

TL;DR: If a brain (computer) is going to be capable of getting off the planet before it is destroyed, it is going to need to develop spherically so all the components of that brain function to its greatest capability. In order for that to happen, it must be supported from the base of its skull – as though it were suspended on a pole. For that to occur, a spine would be needed to support it in that way. An ocean plus a billion years of horizontal evolution would be the perfect setup for this development to begin.

As the fungus began iterating through various configurations of MCCTs, a new branch of organisms formed that were much larger in size and capable of moving through the aquatic environment with more grace than their predecessors. These early ACTs possessed soft bodies and existed near the floor of the oceans, akin to the jellyfish of today. These creatures were much more successful than prior life forms at moving around and gathering nutrients, which allowed them to continue to evolve in a variety of different directions.

The most important mutations that occurred in these early ACTs were those of a bony structure. The earliest ACTs that possessed some form of rigid bony structure were fish that housed this bony material on the exoskeleton. Further evolution pushed these bony structures inward, generating the first vertebrates. These creatures were more rigid than their jelly ACT ancestors but more flexible than their bony exoskeleton ancestors. This balance between flexibility and rigidity was pivotal for striking the proper balance between durability and mobility. These ACTs are the ancestors of the fish we know and see today.

As aquatic life exploded and competition became fierce, some ACTs began to try and escape the harsh life and highly competitive conditions that plagued existence in the oceans and other aquatic environments. This forced these ACTs toward the shores of their aquatic environments. In doing so, evolution continued and amphibians appeared on the scene. An example is the modern tadpole, which develops both kinds of organs needed for survival in aquatic life as well as life on land.

"Halfway through its development a tadpole has both lungs that breathe air and gills for breathing water." – William A. Whitelaw, Professor at the University of Calgary, *Scientific American.*[51]

Another example of escaping evolution is the flying fish. If this fish did not develop its fins into wings, it would have gone extinct. Today this fish can fly as far as 650 feet, long enough to escape its predator.

"Flying fish are thought to have evolved this remarkable gliding ability to escape predators, of which they have many. Their pursuers include mackerel, tuna, swordfish, marlin, and other larger fish. For their sustenance, flying fish feed on a variety of foods, including plankton."[52]

By adapting to both environments, these amphibians gained access to a greater level of energy due to the increased levels of oxygen in the atmosphere when compared to water-based environments. The fungus was preparing to make its journey onto land and begin the next phase of organic evolution. This also is connected with the development of a brain capable of higher levels of cognition, memory, and consciousness due to the greater amounts of energy afforded by the increased concentration of oxygen in the air.

Fungus' Primary Goal

In this stage of the fungus, the vast array of permutations and combinations that were made possible by moving toward multicellular life begin to take on new forms and functions that become refined and honed to efficiently survive the aquatic environments in which life continues to thrive. This resulted in the development of organisms that were well-suited to move throughout the aquatic environment at relatively high speed with specialized appendages or bodily structures. Early ACTs were the first organisms to house primitive brains. Ultimately, this resulted in the development of vertebrates. These adaptations allow for further exploration of the aquatic environment and the eventual emergence onto land, and in doing so, push conscious awareness one step further. The fungus is moving one step closer on its cosmic journey to TES and MCG.

EOS

- Head and spine development.
- Ocean food sources evolve.
- The need to leave or evacuate the ocean is eventual for some species to survive.
- Only on land can a brain eventually become upright for its spherical development.

QCT - The Erectors

TL;DR: The *fungus*' intuitive growth is innate in that it must survive on all fours mastering the land by adapting to a new predator/prey environment. Developing the upright posture allows for the growth of a brain that will succeed at TES. It will mutate from several different conditions over millions of years, even going through mass extinctions. It's roaming on all fours terrestrially that is the honing system for the fungus brain growth and the preparation for bipedalism.

As each of these transportive organisms had been evolving beyond the MCCT stage, the split of fungi from plant and animal life 1.3 BYA had begun the preparation of food on land for the eventual exodus out of the water that would eventually occur. As the earliest amphibious creatures found themselves existing for increasingly long periods of time on land, the abundance of plant life that had developed would support their continued evolution.

These ACT vertebrates marked a key development in the evolution of the fungus, as their more rigid body structure allowed for the transition from the aquatic environment to the land. With more exposure to land, these amphibious ACTs and tetrapods evolved further until their existence on land became permanent and the QCT was formed. The development of lungs to respire proved to be more efficient in terms of extracting oxygen from the air in comparison to the gills of the ACTs. Early QCTs were not warm-blooded; however, with mammalian evolution, more energy was required in order to sustain proper thermal regulation. One way in which these mammals were able to offset these energy requirements was to eat more frequently. Luckily, around the same time QCTs evolved, plant life had been flourishing on Earth. These plants served as nutrient-rich sources of energy for these newly evolved QCTs. In addition, some mammals were also carnivorous, so their energy requirements could be met by eating other QCTs and ACTs. These factors allowed the QCTs to develop larger brains than most of the ACTs. Since QCTs walked on four legs, of which the rear two evolved from fins, they were mainly adapted to moving around the landscape. Interestingly, some QCTs were forced back into the waters because of climate change due to volcanic activity during the Permian extinction. A large portion of life went extinct both on land and in water.

"The process of evolution has been identified in dolphins and it is believed they are connected to many terrestrial mammals. This is based on research that shows them to be part of the Artiodactyl order. What has been identified are ancestors of the dolphins that date back more than 50 million years. Before you say that there is no way that dolphins are related to land mammals, you have to look closely at the facts."[53]

These creatures were forced to adapt their physiology based on their proximity to bodies of water in order to survive the stressors of predation and changing climate. The interesting thing about this particular evolutionary change is that, for the first time, a purely mammalian form of respiration was preserved in the aquatic ecosystem. Essentially, the form of respiration that was learned by the fungus on land environments was utilized in the environment in which the *fungus* originally had not iterated such a solution or feature of organic life.

"The theory is that some land-living ungulates favoured munching on plants at the water's edge which had the added advantage of allowing them to easily hide from danger in shallow water. Over time their descendants spent more and more time in the water and their bodies became adapted for swimming. Their front legs became flippers and a thick layer of fat called blubber replaced their fur coats to keep them warm and streamlined. Eventually, their tails became bigger and stronger for powerful swimming and their back legs shrunk. Gradually, their nostrils moved to the top of their heads so that they could breathe easily without the need to tilt their heads while swimming. As some of these creatures began to feed on a different diet, they evolved into baleen filter feeders and lost their teeth."[54]

As further evidence that land mammals crawled back into the ocean over geologic timescales, the spinal structures of whales and dolphins more closely represent those of mammals instead of fish. After becoming mammals they developed back into flippers as marine mammals. For example, fish have spines that move horizontally but not dolphins. Instead, they move vertically, which is the same as animals that live on land.

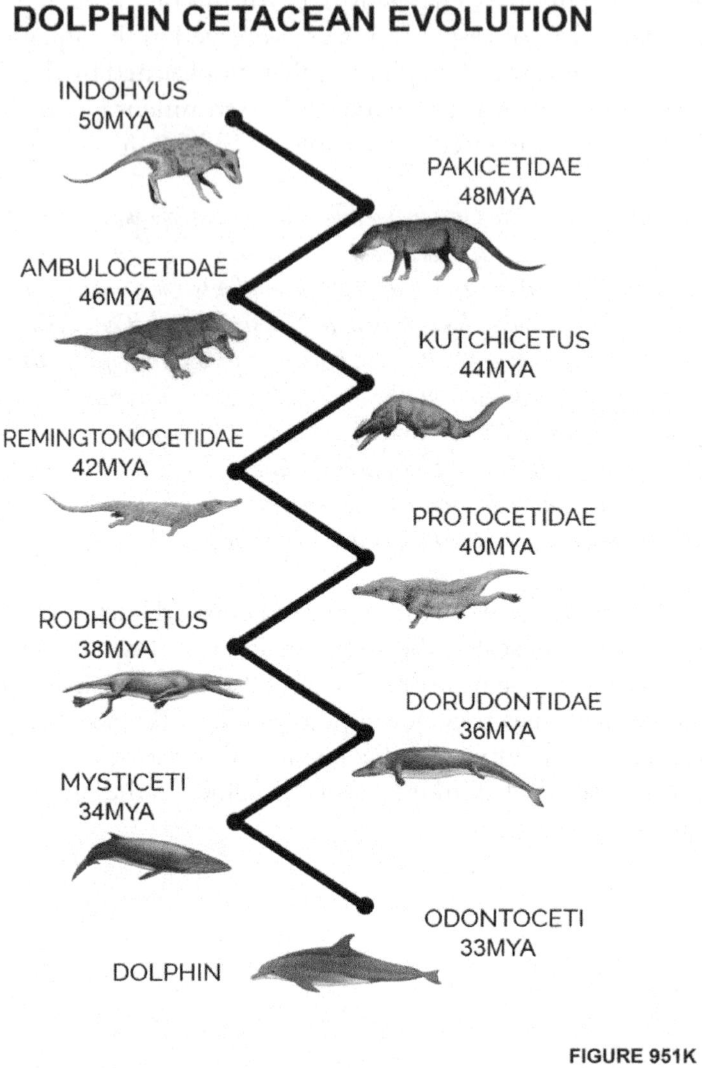

Figure 951K (Dolphin Evolution)

This figure demonstrates that evolution is dependent on evolving into one environment or back to a previous environment in order to survive. In the case of the dolphin, it survived by slowly evolving back into the ocean.

It becomes much more probable that whales and dolphins have these anatomical structures because they basically started out this way as land mammals. Since these aquatic mammals respire oxygen through their lungs, the increased efficiency in oxygen uptake compared to fish allows them to develop more complex brains. The intelligence of these aquatic mammals rivals most other life in the ocean, with dolphins and whales displaying very complex behaviors and modes of communication. The whale and the dolphin are evidence that the increased use of oxygen is inevitable for the development of increased intelligence; otherwise these land mammals would have evolved to using gills once they entered the water again.

While archaeological records have been the gold standard in developing phylogenetic trees to describe the history of organisms on Earth for the better part of a century, this methodology has recently come under scrutiny as having the potential for producing incorrect associations or lineages. The idea that bone structures show clear evolution, that is that one bone structure comes after another, has been suspect as independent genetic mutations can occur at different times in different groups of animals. This particular issue, carried forward for millions to billions of years, could create erroneous connections and orderings in the evolutionary trees we have currently come to conclude are the logical order in which life evolved on Earth. In order to ensure that these types of mistakes are cleansed from the phylogenetic trees, a better method for discovering the evolution of species is needed. Recently, genetic analysis has become robust enough to act as such a method. This will be discussed later on in the chapter; however, it is sufficient to say that the mentioned migration from land to sea actually did take place in the manner described.

At one point, a very important evolutionary step was made in Jurassic reptilians. One such reptile, the Monkeydactyl (~160 MYA) paved the way for organisms to manipulate things in a technical manner. This evolutionary trait became one of the most significant changes to animal life that would allow for greatly enhanced manipulation, particularly in a technical manner, of an animal's surroundings, as well as dexterity in general. It wasn't until this feature was found in early primates that it began to demonstrate its importance.

With the opposable thumb, early primates began to evolve into animals that were masters of the canopy. These evolved primates contained a variety of species, but the most important were the apes. Swinging from limb to limb and being able to manipulate fruit and other food, these animals evolved further and further. With changes to the climate, trees became more scarce due to warming on Earth. These apes were forced to move farther and farther over exposed land before making it to the next tree, which put them at risk of becoming prey to other predators in the process. This gave an advantage to those apes with the ability to stand further upright and to do so for longer periods of time than their peers. This selective pressure forced these early apes to become more upright in stature until the very first hominids evolved. The first human, sharing greater than 99.6% similarity in its DNA to that of a modern human, demonstrates just how incremental these changes are in bringing about the evolution of one species to another.

Fungus' Primary Goal

During the evolution of the QCT, the *fungus* iterates through many refinements. Most notable have been the increased efficiency of oxygen uptake through the use of lungs, the evolution of larger brains, the automatic maintenance of internal temperatures (warm-blooded), and the appearance of the opposable thumb. These advances brought the fungus to a stage where it had arrived at an organism capable of an unprecedented.

BCT - The Evacuators/Spore

TL;DR: It is at this point that the *fungus* can finally get the brain capacity large enough to look upwards and become conscious of its total environment. By understanding the significance of water, land, sky and space, the capacity needed to survive its own planetary demise. Now that the fungus has reached the level of significant conscious-growth, it can make reality of its inevitable future. It can now go to work at getting the hell out. This is necessary if the BCT (spore stage of the fungus) can ever evacuate its inevitably doomed birth planet. This is a slow and meticulous change through time so the BCT's consciousness can barely be noticeable to itself.

Now that the *fungus* had iterated its way to the BCT (early hominids), the journey toward the spore phase (refer to the comparison of evolution to the "Mushroom Life Cycle " in Chapter 2) nears completion. Early hominids were the first animals to walk on two legs, albeit not fully erect. These early hominids crafted tools from bones in order to more effectively hunt. By having the ability to use two arms with hands that had opposable thumbs, two feet that could allow them to walk and run, and retaining some of the climbing ability of their ape ancestors, these hominids could traverse any landscape. For this reason, they were capable of finding food sources in all different regions of the Earth: land, water, and even the sky (birds, insects and rain). This increase in the availability of food allowed for continued brain growth, which resulted in the further evolution of the BCT.

These hominids evolved into the first ancestors of modern humans. These successors to the hominids were able to stand fully upright and continued to evolve larger brains than hominids. These larger brains were also allowed to grow more easily in the cranial cavity of upright BCTs as the increase in size has much less of an effect on the balance of the organism when compared to QCTs. A QCT with an increasingly larger brain at the end of a horizontal spine will have its center of balance shifted forward, making them top-heavy. Similarly, the fact that the front legs are liberated and become arms, which frees up the jaw from having to perform most tasks, allows for infinitely more interaction with the environment, which stimulates brain development and growth. These factors coupled together make the BCT a "perfect storm" of evolutionary changes allowing for an organism that had unprecedented levels of evolutionary potential.

While Homo erectus was the first to stand completely upright, it wasn't until Homo sapiens that humans began revolutionizing the technologies that had been used by their ancestors, starting with Homo habilis, millions of years prior. It was at this point that the technical revolution of the fungus had begun. Given the increased ability for brain evolution and adaptation afforded by two free hands, each successive iteration of humans became more capable, more intelligent, and more resourceful. While early technologies took tens of thousands of years to show signs of change, this is to be expected in an exponential process. The beginning growth of any exponential trend starts relatively slowly; however, after a certain point, the inflection changes significantly and growth occurs rapidly. To demonstrate this fact, stone tools that were developed by early humans were used for millions of years before any metallic tools were developed and utilized in the Bronze Age. Then, the Bronze Age only lasted approximately 2,000 years. Consider that now, in modern times, we have gone through multiple ages within a 100-year span. The BCT is truly a marvel of the fungus and will continue to develop as the spore of that fungus in order to ensure that TES are achieved for the survival of that fungus as a TCM. Should humanity make it to this point, it will have evolved so much that it would be considered a GCT, as the primary mode of transportation of that organism would be entirely artificial. A human at this stage of evolution is transported in an entirely artificial environment throughout the galaxy and potentially other galaxies. While these humans may still utilize their legs for transportation around their vessels, or on planets they may visit, this mode of transportation becomes seemingly insignificant in contrast to their existence throughout a much grander environment: the cosmos. Humans have started becoming smaller as the brain has evolved because their physical strength is not required due to the advent of technologies that have assisted them. Continuing this trend, the evolution of the BCT will likely show up in the GCT as an enlarged brain and a reduced physical stature.

"As we have acclimated to warmer temperatures, the way we live has also generally become less physically demanding, which overall serves to drive down body weights."[55]

Evidence of Mankind's Evolution

Humans today still have semi-movable ears, which is a remnant of behaviors that animals used to locate predators or prey when in their environment. Being able to move the ears allows for better perception of the sound as well as a greater ability to locate objects by paying attention to the amplitude of the sound as the ears are moved. Humans also have a tailbone but no tail; however, if you look far back enough in the evolutionary tree, our ancestor mammals did have a tail. Some humans are born with no wisdom teeth, showing evolutionary changes that are occurring in modern times. This is consistent with the archaeological findings that demonstrate the human jaw began to shorten around 1.8 million years ago. Similarly, a bone spur has begun to show up on the back of millennials' skulls near the base where the neck meets the skull, called the occipital bone, which is being attributed to the changing posture and living conditions of humans. Recent articles by the BBC and *The Washington Post* have cited a 2018 study in the journal *Scientific Reports* saying that these bone growths have been turning up more often than expected in people aged 18 to 30. The study suggests that "sustained aberrant postures associated with the emergence and extensive use of hand-held contemporary technologies, such as smartphones and tablets" are to blame.[56]

Another way one can see evidence of human evolution is through embryology, the study of human embryos as they develop in the womb. Embryos of many different kinds of animals-mammals, birds, reptiles, fish, etc. look very similar and it is often difficult to tell them apart during their embryonic development. Many traits of one type of animal appear in the embryo of another type of animal. For example, fish embryos and human embryos both have slits. In fish, they develop into gills and the jaw but in humans, they only develop into the structures of the jaw. Human embryos also display webbed structures in their hands and feet, which resemble the webbed features of amphibians and some birds.

"Human embryos resemble those of many other species because all animals carry very ancient genes. These genes date back to the origin of cells, which are expressed during a middle phase of embryonic development."[57]

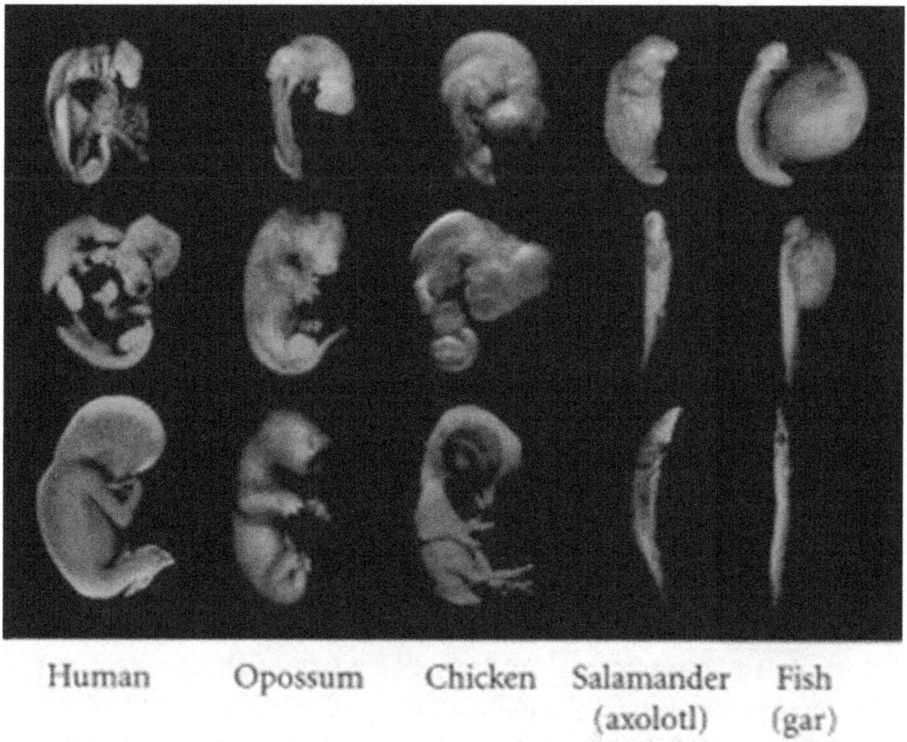

Figure X.XC: Phases of Embryonic Development

This image, which comes from a study performed in 1998 by M. Richardson (Haeckel, embryos, and evolution), demonstrates that many different animals go through phases of embryonic development that look quite similar to one another. The genetics that controls each of these creatures' embryonic development are passed down from ancient ancestors, with differing evolutionary changes here and there that create the unique organism in each case. What is apparent is that these early stages each represent a common point where evolution generated a template from which many different permutations and iterations took place to generate uniqueness in the vast multitude of organisms on Earth.

All humans form into females first and it isn't until the production of certain hormones that the sex organs are rearranged into that of a male when males are born.

All of these examples demonstrate that this process of development is also fungal-like in nature. They also demonstrate that humans are part of an evolutionary process that continues to this day. This is direct evidence that animals are a part of the human evolutionary process as can be seen when watching a sort of "time-lapse" movie of that evolution as the human embryo develops. Many traits are carried over and others that are unnecessary fade away. However, the end result of going through the permutations of what was needed, what was unneeded, and what was most advantageous is the ability to be seen through this embryonic history.

There are also evolutionary changes in humans when compared to animals that give humans uniquely distinct advantages over all other animals that came before. In newborn humans, the skull is flat and the voice box is raised, which makes it impossible for the newborn to produce complex language until later when the voice box drops down to the location where it resides in adults. This shows the evolutionary changes in humans that gave them unique advantages that promote increased intelligence through communication and expression.

"...In the course of that evolution, the larynx developed and acquired the ability to vibrate. In humans, the larynx descended in the neck, resulting in an effective articulatory system, the ability to communicate through speech was achieved..."[58]

The importance of the development of larynx in human evolution is affirmed by a study published in the journal *Quality in Primary Care* in March 2017.

"In the course of evolution, the larynx developed and the acquired the ability to vibrate. In humans, the larynx descended in the neck, resulting in an effective articulatory system, the ability to communicate through speech was achieved ... Laryngeal elevation and glottal closure are indispensable for human beings."[59]

"A high larynx, says Laitman, also suggests that Neanderthals couldn't make the

same range of sounds that we can, since a lower larynx allows for a larger sound-modifying airspace above it.'[60]

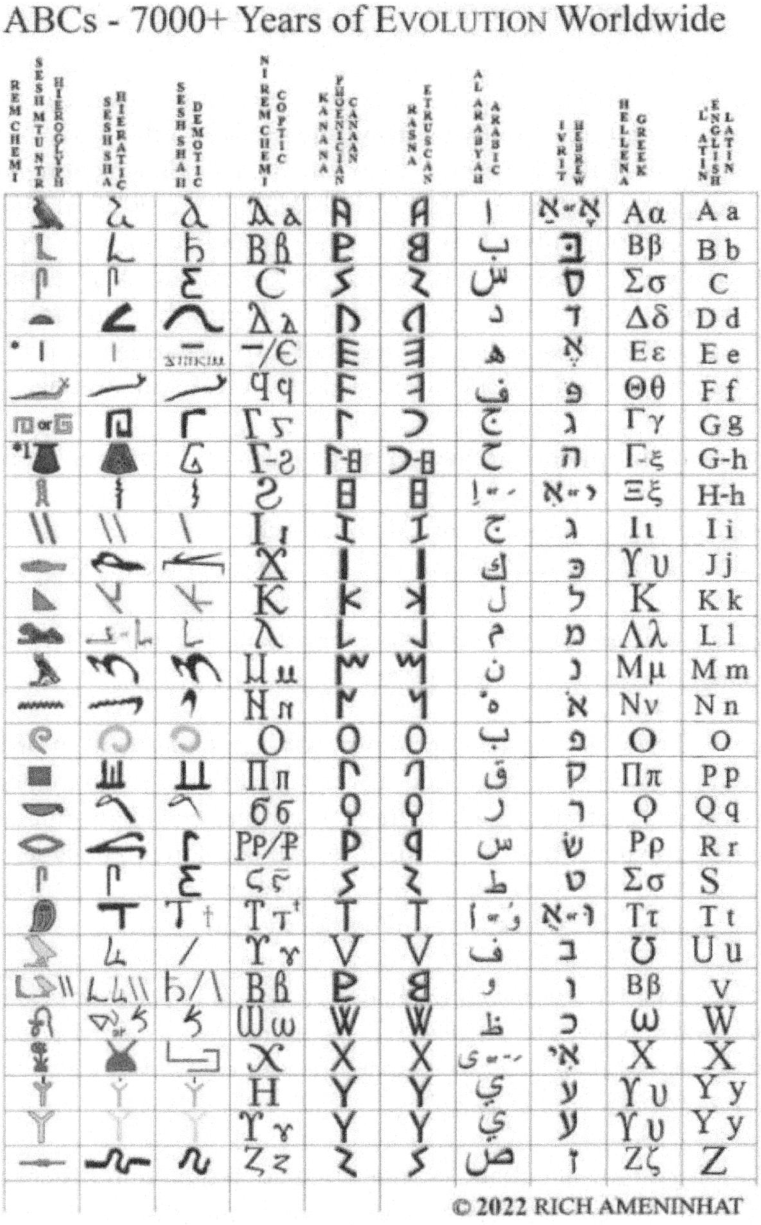

Figure BK11: The Evolution of the Alphabet.

This figure shows how conscious growth is not possible without the physiological evolution of the modern human. With the need for more and faster communication between each other every evolutionary year, the vowel and consonant exchange become more flexible. This is directly related to the larynx dropping down to its optimum placing, allowing for the development of sounds that make the need for the scribing of them for re-interpretation and faster information via vocal cords.

There are many other examples that could be cited that describe the progressive evolution of humans from one species to the next; however, the ones that have been mentioned are evidence enough that this process has occurred from the very beginning and is continuing to occur, even in modern times. By analyzing the archaeological records and denoting the changes that have occurred in human skeletal and anatomical structures, when compared to both prior humans and other animals, as well as considering the needs of a space-faring civilization, some predictions can be made about what the GCT will look like.

How to Stay in Space Forever - The process of Transhumanism

One obvious change that has already been occurring in humans is the shortening of the jaw and the shrinking/receding of teeth.

"Shrinking jaw muscles may have helped a gorillalike skull (right) evolve into a human one, making way for a bigger brain."[61]

Humans will rely less and less upon powerful jaws and teeth in order to provide sustenance as hunting and tearing apart flesh are no longer the primary modality of ingesting nutrition. With further enhancements in technology, the usefulness of strong jaws will continue to decline, as well as the need for sharp teeth. This is evident in that early man did not have crooked bottom teeth, only modern man.

"Over time, as jaws became smaller in response to these dietary changes, teeth didn't follow suit, remaining around the same size. This likely directly led to the problem of adequate mouth real estate so common today. We've got modern jaws but a potentially outdated number of teeth."[62]

An example of more dramatic changes that are required in order for a human to survive solely in space could be something we've already seen before. Eye-witness accounts of abductees indicate that many, if not most, people claim to have seen beings with slender bodies with large eyes and heads. These accounts seem to coincide with the logic of what would be required for a being to be suited to survive mostly in space. The fact that many people in the same incident are able to recount very similar details without being held in the same room during their interviews is also extremely telling. Humans that exist entirely in space will not be required to lift heavy objects or utilize their strength to run long distances. Due to the reduced physical requirements of permanent life in space, the human anatomy will begin to shift in order to make the most efficient use of its energy. Given that intelligence should increase and physical strength will no longer be as important, most of this energy will be used by the enlarged brain.

However, not all changes to human physiology will occur simply because their environment has changed. Some of these changes will need to be forced to occur before the environmental changes take place. Future humans can make these changes much more easily due to the advent of CRISPR early on in the 21st century. With continued research into the human genome, and further improvements to CRISPR and other gene editing technologies, the ability to create tailored anatomies with unique biochemical capabilities will be commonplace.

Sometimes these improvements to the genome can be instructed by preexisting organisms found in nature. The tardigrade, a microscopic animal with eight legs that is part of a class of organisms called extremophiles, can survive the conditions of the cold temperatures in space, is unaffected by radiation, and evolved 530 million years ago during the Cambrian period. By using the DNA from a tardigrade, humans could potentially incorporate the DNA or modify it in a manner consistent with the human equivalent of the tardigrade DNA using CRISPR in order to become a GCT that would be capable of living without water for a long period of time while resisting the harsh radiation that exists in interplanetary and intergalactic space.

"But in a Weizmann Institute of Science study published today in Cell, researchers have grown synthetic embryo models of mice outside the womb by starting solely with stem cells cultured in a petri dish – that is, without the use of fertilized eggs."[63]

The research opens a window of possibility into a future where reproduction is simpler, faster, and less complex than regular reproduction through mating of sexes.

Another wise choice that would be made by GCTs would be to reduce their reliance on organic sources for energy. Radiation is abundant in planetary systems due to the output of each planetary system's star. Utilizing this radiation as food would allow for an essentially limitless supply of energy for future humans to thrive upon. This is not too far-fetched an idea when one considers that humans already utilize the sun's radiation in order to generate vitamin D as these rays are absorbed through the skin. Ultimately, genetic engineering that reshapes humanity into something vastly different from what it is today would be required to create a GCT.

Reproduction would also need to be totally controlled by the GCT, in both quantity and quality. This could mean cloning is a potential option, where all of the processes for reproduction are housed externally from the GCT. Another possibility is that the GCT could mate with itself, like hermaphrodites (e.g. earthworms). Cloning will eliminate the primal function of procreation, potentially giving way to advances in GCT brain development in comparison to the random advances made currently through sexual reproduction in BCTs. This form of advancement will accelerate the cognitive abilities necessary for surviving Earth's end and beyond.

Fungus' Primary Goal

TL;DR Transhumanism is a philosophical and social movement advocating for the use of technology and science to enhance human physical, intellectual, and psychological capacities. It seeks to transcend the natural limitations of the human body, such as aging, disease, and mortality, through advancements like genetic engineering, artificial intelligence, and cybernetics. The goal of transhumanism is to improve the human condition and potentially evolve beyond our current biological state into a "posthuman" future to survive in outer space, the fungus subconscious making the first step in building a GCT.

By the time the fungus has gone through enough iterations to develop a BCT, the main goal of the fungus becomes more polarized than before. Initially, the fungus has been iterating in order to produce sufficient complexity and adaptability in organisms in order to generate a level of consciousness and intelligence that can create technologies. At this point, the focus shifts toward refining that organism further, as it is all that is required to ensure that TES are achieved. As the BCTs become more intelligent, more of the collective consciousness of the BCT becomes polarized and acutely aware of their place in the cosmos, their limitations in their star system, and what this means for their continued existence. With the exponential nature of technological growth, the BCT will begin focusing more and more resources toward achieving ESS to ensure that life as a TCM is successful. The ultimate goal is to produce a GCT that will go on to evolve and spread throughout the cosmos in order to reach MCG. The more these BCTs spread and evolve, the closer they get to reaching MCG and the UGCH.

Going Beyond Darwin

TL;DR: Whether it be evolution or creation, it is important to keep the truth from surfacing into reality until it is a needed realization for the species to survive evacuation day. Darwin's dedicated work in finding the evidence that all animals have evolved from some common ancient species is the best evidence of system (EOS) we have to date. Finding out the physics behind the physical plane in which this system may evolve (a biosphere) supports Darwin's work as a system necessary to survive by evolving into an animal that can adapt to post-terrestrial existence.

As was mentioned prior, a less error-prone method is required to ensure that the phylogenetic trees that have been created by analyzing fossil records are accurate. This method, the analysis of genetic variation and the associated changes from one species to another, can confirm or deny whether the order in which species have been said to evolve, as well as the relationships between ancestor and successor, is correct for myriad groups of organisms on Earth. To demonstrate, with much more solid evidence, that the transition from land back to water actually occurred, genetic analysis of multiple fossils has confirmed that this did actually happen.

"Fifty million years ago, the ancient ancestors of whales and dolphins roamed the land on four legs. But over time, these aquatic mammals have evolved to live fully in the ocean—their genetic makeup changing along the way. Now, a group of scientists have investigated the changes in 85 different genes that were lost in this land-to-sea transition."[64]

This is a particularly important evolutionary step in the evolution of life on Earth because it demonstrates that evolution likely occurs more than just at random. If random evolutionary changes were the only changes that drive evolution, it would be unlikely that the exact changes required would always show up in so many organisms just in time for them to survive their changing environment. Thus, it can be speculated that there is more than mere random variation in genetics at play; there is an exchange between the environment and the organism that is more than random, which drives evolution. While this process may occur outside of the conscious control of most organisms, we have clues that this process does occur at a conscious level in some organisms, like the octopus.

"Unlike other animals, cephalopods – the family that includes octopuses, squid and cuttlefish – do not obey the commands of their DNA to the letter. Instead, they sometimes interfere with the code as it is being carried by a molecular "messenger." This has the effect of diversifying the proteins their cells can produce, leading to some interesting variations."[65]

While octopi are some of the only creatures we have found to date that have this particular capability, it opens the door for other lesser interactions likely occurring within the organism but at a conscious level. For example, organisms are constantly trying to maintain homeostasis. While they do this, all systems generally adhere to least action principles. That is to say that the organismal machinery and chemical pathways will adhere to physical laws and processes that utilize the least amount of energy required in order to come to completion. In doing so, with enough repetition, these processes and influences are "priced in."

For example, if a human eats tons and tons of sugar, it will eventually reduce the sensitivity of its cells to insulin. This is to keep the massive amounts of insulin required to bring the blood sugar down from doing so to such an extreme that the blood sugar will drop to a dangerous level, resulting in the death of the organism. Many such chemical and environmental reactions occur throughout the day for organisms. Therefore, the interaction of the environment with the organism imparts non-random changes to the epigenetic (non-gene related changes to the genetic expression in the organism) makeup of the organism. This particular process removes the notion that all evolution is selected due to random variations in genetics. Darwinian evolution and the environment "selecting" the ideal and fittest creatures due to randomizations/variations in DNA may be a part of the evolutionary process. However, it is likely not the primary driving factor in the evolution of these organisms. It is the communication of the environment to the creature through epigenetic factors, and the attempt to maintain constant homeostasis, that is driving transcriptional changes at the DNA level minute by minute. This promotes the evolution of organisms in a much more direct fashion. This would allow and account for why so many organisms seem to be able to survive/adapt to their environment within a sufficient time to ensure that they do not go extinct. If evolution were a purely random variation of genetics, the failure rate of life would be far higher and there would be vastly more phylogenetic trees that end abruptly, being unable to persist through multiple variations and changes in climate.

This also supports the concept of efficiency being a primary motivating factor for the evolution of organisms. As the organism responds to its environment in order to achieve homeostasis, it does so in a way that it interacts with its environment more efficiently. Similarly, structures that are better suited to operating at a more efficient level will be promoted by the environment and if it changes. Thus, an increasingly complex brain that completes more complicated tasks and "computation" but with greater efficiency is directly in line with an organism that is interacting with its environment and evolving through this exchange. The inevitable result will be a creature that is bipedal because bipedalism is more efficient and more versatile than most other forms of life. The trade-off between any loss in direct efficiency of locomotion is outweighed by the diversity (and thus efficiency) of its ability to interact with its environment through the use of arms and hands instead of jaws and claws. In this sense, evolution is a process that may iterate in different ways toward a specific end goal; however, that end goal is constrained in some sense. Evolution would never end with quadrupeds as the most efficient form of life, given enough time. Thus, evolution can be seen as the interplay between information/memory (through genetic transcription) and the environment, which drives a species toward a form that is increasingly efficient in its interaction with the environment. Most importantly, this process is not entirely random, as it is guided by the very interactions these organisms have with their environment at the macroscopic and microscopic/cellular levels.

Evolutionary Fungal Network (Proof of Fungal Development)

TL;DR: If any developing *fungus* throughout the universe was to be successful with its eventual biospheric evacuation, it would need to rely on the mineral and chemical composition from the initial gravitational bombardment. This would give each and every biosphere its own unique set of element- and mineral-feeding bacteria, which is the best way to start that unique cell structure for a *fungus* starter. This is how the *fungus* uses time the host star allows for the development of a BCT or bipedal brain that is capable of computing the necessary outcome and technology for evacuating its birth biosphere.

It is very instructive to look at the development of life on Earth at the grandest of scales. By observing the general progression of the evolutionary process in a manner similar to what a slime mold would look like as it spreads out searching for food, one gains an appreciation for the fungal networking structure that the evolution of life displays.

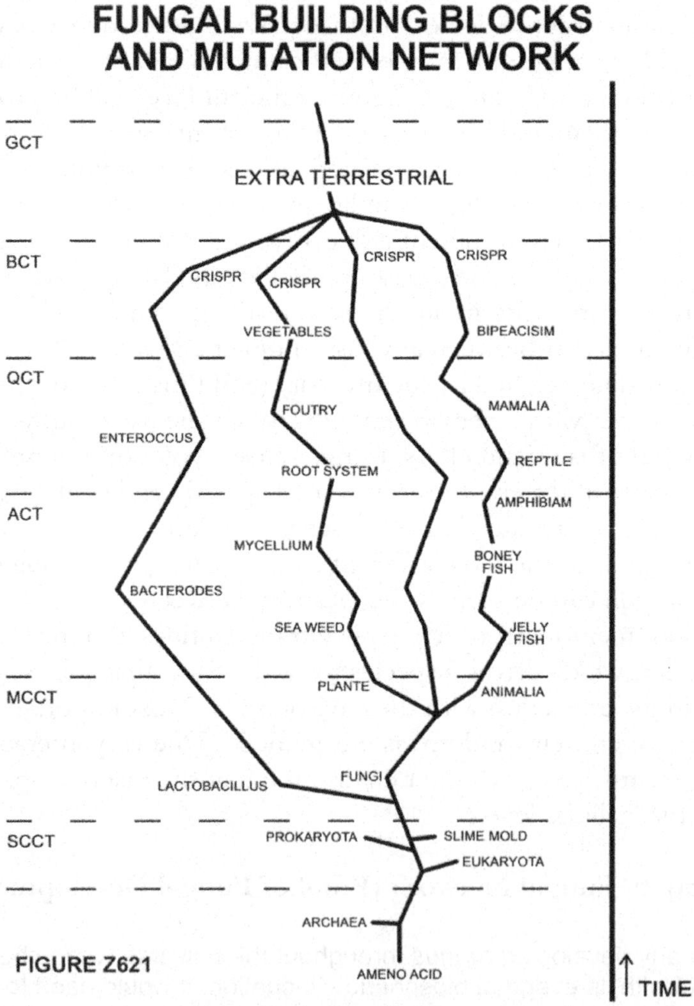

Figure Z621 fungal mutation network

This figure shows the work of Fungal mutation to BCT/GCT, demonstrating the branching nature of evolution through mutations from single-celled organisms to the GCT, which very much resembles a meandering slime mold.

The evolutionary progression of organic life throughout the history of the Earth demonstrates that all life on this planet, constantly evolving, constitutes a living entity, the *fungus*, that strives towards greater adaptability and efficiency. The ultimate goal of this *fungus* is to produce a spore, the GCT, capable of supporting itself post-terrestrially through ESS. This is the case because each of the individual organisms that evolved have done so in a manner that supports the whole by adapting to niche environments of both energy absorption/expenditure and raw material processing. Each organism depends on some other(s) and each one supports some other(s) in such a fashion that the entire ecosystem works together like the moving gears of a clock. Yet, while this is true, it is extremely difficult to say which is more/less important; or equivalently, which organism promotes the existence of another. For example, it has been said that without the seemingly insignificant bee, the entire world would run out of food in less than half of a decade. However, without flowers, there would be no bees. This small example demonstrates just how vastly interconnected the web of life is without having to address more than two types of organisms. The task of imagining just how much more complex and interconnected it is will only be limited by the imagination of the ponderer. Surely, the limit of this connectivity is beyond the imagination of any human as it must preserve HTO.

Though some of these paths may never lead to bipedalism, they are all necessary and utilized by the *fungus* in some way to achieve it. The species that are members of these paths also serve to explore the "possibility space" that allows optimal conscious-growth to be obtained by other paths of the *fungus*, whether this be through indirect support (e.g. serving as food for another pathway of evolution) or direct support (e.g. by promoting environmental changes that influence species that do contribute to the development of a BCT). Another such method is horizontal gene transfer, when genes from one organism are passed to a completely different one through a variety of processes (e.g. viruses transferring genes from one organism to another). Regardless, and as has been stated before, the ideal path would not be realized as it would not have been explored if these seemingly unsuccessful organisms had not evolved. This would make each branch a significant part of the stability, preparation, and development of what would eventually become bipedal and, ultimately, a BCT/GCT to exist as a TCM. This demonstrates how finely tuned and interconnected yet resilient this entire organism is. While some species do go extinct, this does not detract from this point. This merely reinforces that there are high degrees of tolerance at some levels and high degrees of reliance at others.

This is directly in line with the principle of HTO, as it is nowhere near obvious which creatures could crumble the house of cards should they be removed. It is also incredibly hard for any developing consciousness to trace back its origins through the complicated web that the *fungus* weaves. The self-feeding nature of the *fungus* makes it even more difficult to determine which ones are truly fundamental organisms, as it would seem that everything relies upon something else.

Carrying this idea forward and extending it beyond the scope of Earth, over billions of years, a *fungus* utilizes the energy of its system's host star and grows in size and complexity as it iterates through various possibilities in the forms of various organisms until it has iterated enough to produce a BCT. By this point, viewing that particular *fungus'* evolutionary tree would demonstrate the same networking structure as exists here on Earth, albeit different organisms and branchings would exist. Subsequently, that BCT would continue to evolve to achieve TES and would go on to exist in the cosmos as a TCM if the *fungus* that biosphere produced is not a dud.

Evolution of the Brain and its Relation to Conscious-Growth

TL;DR: The first single-cell reaction to starvation is one that takes a certain amount of calculating in order to feed off minerals or one another. This is the first behavior that resembles computer or brain activity that must have been present in the very first cell in order to divide and ignite life on Earth. This continues for eons, building a consciousness until the *fungus* brain looks for food elsewhere other than its host planet.

"Computer modeling revealed that tension and strain on the fibers dictated which set pattern of cirri positions was possible at each moment. Some cirri store stress at different stages of a gait; when that stress is released, it propels the cell to move forward into the next state, causing a cyclic transition between these states. "The fact that Euplotes' appendages are moving from one state to another in a non-random way means this system is like a rudimentary computer," says Marshall."[66]

Since we have shown the similarities between an actual slime mold and the evolutionary forks and splits that have occurred in Earth's *fungus* throughout its evolution, it would only be prudent to question what Earth's *fungus* is growing "towards" if slime molds grow the way they do in order to approach food sources. The answer can be seen when one analyzes the changes in the structures that house information, memories, and imbue greater intelligence upon the organisms that house these structures.

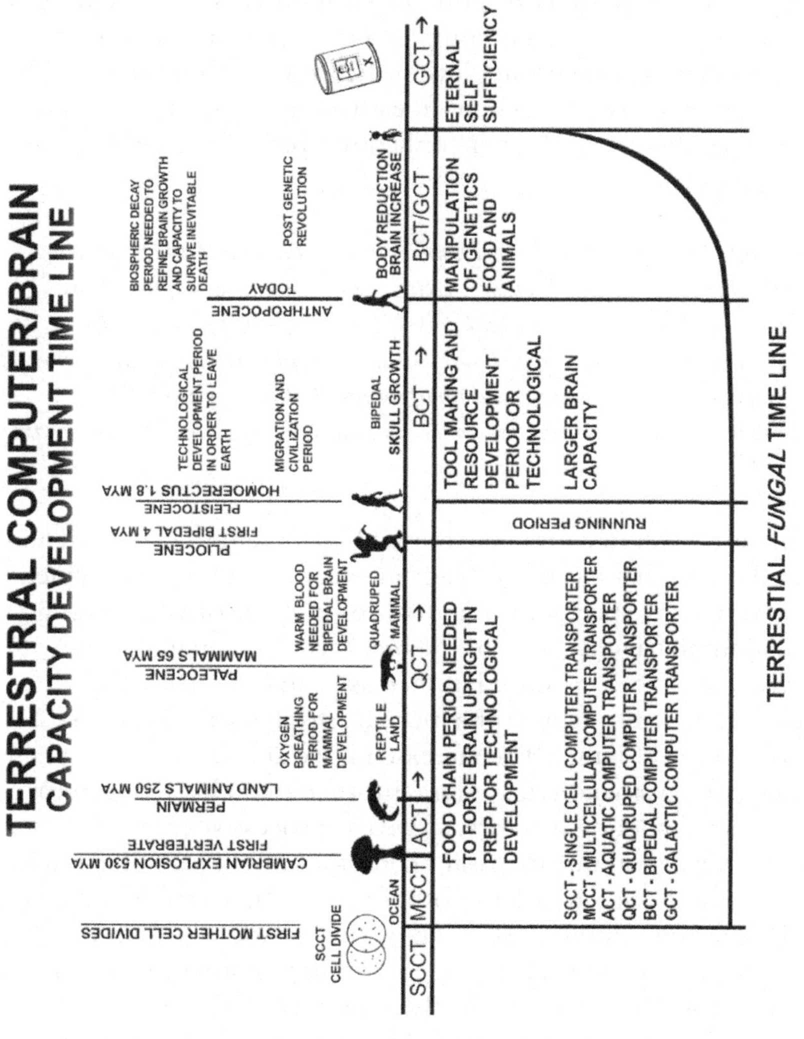

Figure 933Q

This figure demonstrates that throughout the evolution of the fungus, increased complexity and development of specialized structures resulted in the formation of a brain. Though the SCCTs and early MCCTs did not have this structure, once this structure formed in later MCCTs and ACTs, it continued to develop and grow in size ever since. This demonstrates how important the brain structure is in terms of pushing the fungus forward toward achieving TES. The proposed GCT brain continues this trend of increased brain size, while also hypothesizing the physically reduced stature of the GCT, since brute strength is no longer necessary for an advanced BCT.

As a slime mold is growing forward toward food sources in order to grow and ultimately produce a spore for reproduction, the fungus is growing toward an increasingly complex structure that can house and process information with greater levels of complexity and speed, with the same goal of spore production. In early organisms, such as single-celled bacteria, there was no brain, only tiny layers of protein fibers that form the cell memory.

"The fiber ends that pass through the cytoplasmic membrane into the periplasm interact with a complex layer of sensory receptor domains whose conformations depend on the binding of various stimulatory ligands, including several different amino acids and sugars. The intracellular aspect of the fibrous cortical layer interacts with and controls a protein kinase and phosphatase that release a small phosphoprotein that binds to flagellar motors to induce reversals in rotation, leading to tumbles."[67]

However, as evolution continued, these fibers became a dedicated and specialized organ that has been so advantageous that it has continued to evolve across many species for hundreds of millions of years. During this evolution, the average size of the brain when compared to the overall volume of the organism has increased and the shape has continually changed to represent a more hemispherical appearance (maximizing total volume). The increase in size and shape of the brain has allowed organisms to retain more information in both long-term and short-term/working memory, as well as process streams of incoming information due to greater consciousness. These changes to the brain are necessary in order to produce a species capable of being a spore, since a spore in this sense is an organism readily able to develop technologies that are advanced enough in order to bring the fungus out into the cosmos. More specifically, these increases in memory and processing promote anticipatory behavior, which we will denote as "concern." This concern allows for planning and developing far in advance of what is currently being experienced by the organism. This behavior is impossible without greater and greater levels of long-term memory, as you cannot be concerned about something you don't have memory of. In the case of the BCT, it is the evolution of the left prefrontal cortex that has given rise to the greatest differences in future planning and anticipation.

"The prefrontal cortex is commonly associated with cognitive capacities related to human uniqueness: purposeful actions towards higher-level goals, complex social information processing, introspection, and language...Results reveal different scaling coefficients in the left versus right prefrontal hemisphere, suggest that the primary factor underlying the evolution of primate brain architecture is left hemispheric prefrontal hyperscaling, and indicate that humans are the extreme of a left prefrontal ape specialization in relative white to grey matter volume."[68]

It is all about a very slow brain growth process over millions of years. So slow that what may become conscious enough to realize that process is at work, can't cognitively identify that it is a part of the process itself. It can only be viewed through discovery of fossilized skulls. From the evolution of a basic headed animal and the memory cell that expands from adaptation in its environment, reproduction and mutation will make the next necessary brain leap in capabilities eventually surviving the doomed planet it was conceived on.

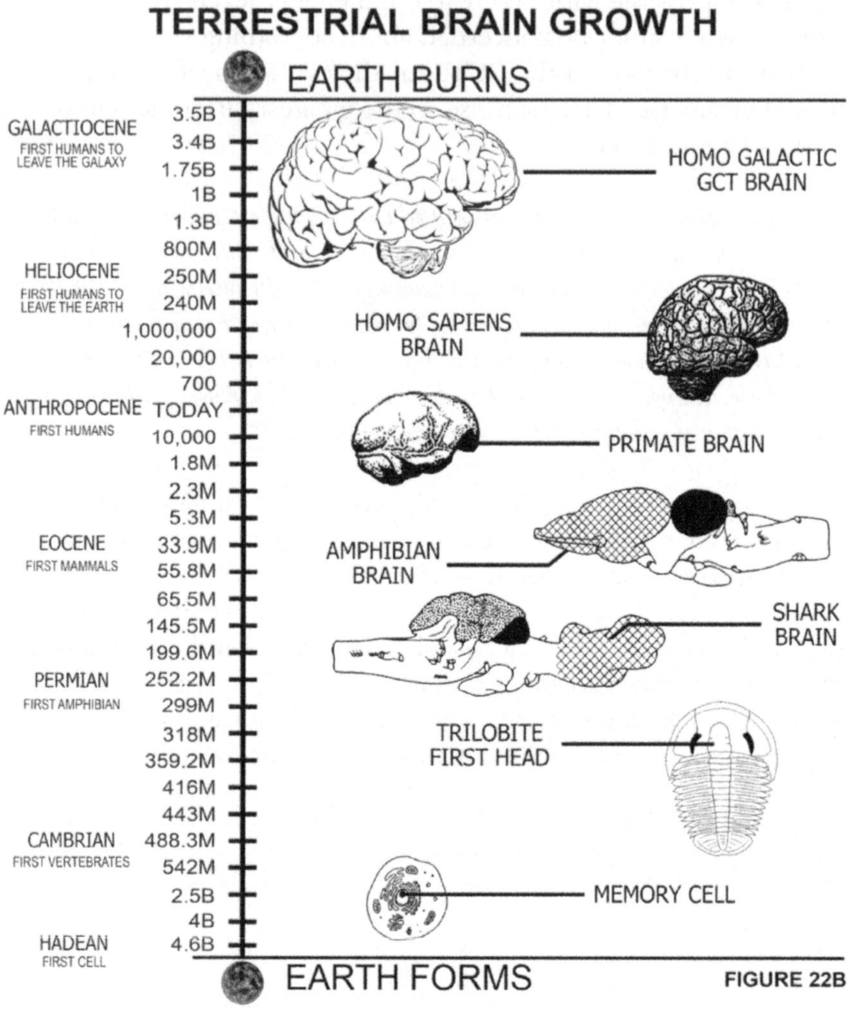

Figure 22B (Terrestrial Brain Growth Process) ticks according to epochs

This figure shows the evolution of consciousness through the development of various organizational structures, which start from single-celled organism capable of possessing limited amounts of memory (memory cell) and evolving toward organisms that display cavities in their anatomy that begin to house specialized processing organs (i.e. the brain). Over the course of evolution, the size of the brain and various structures within the brain grow and allow for increased levels of consciousness and awareness. At some stage, as humanity leaves Earth to exist as a TCM, the brain will grow further in size when compared to modern-day humans; however, the physical stature will have decreased as the necessity to perform manual labor and feats of strength will have been reduced by technological advancement.

The anticipatory behavior of an organism has multiple facets. Dream states are common among most mammals, with other organisms showing similar diurnal cycles in their activity. However, what is unique among humans is the extent to which they can project information forward through simulation. Dreaming is a form of simulation that is seemingly random, so it doesn't represent any structured thought process most of the time. Periods between dreams promote the reconciliation of memories, or their proto-engrams, that form during the day.

"Neuroscientists had long thought that as the day's experiences replayed in the form of nerve impulses during sleep, those memories slowly transferred out of the hippocampus and to the cortex so that the brain could extract information to create rules about the world. They also knew that some rules were synthesized by the cortex more quickly, but existing models couldn't explain how this happened. Recently, though, researchers have used optogenetic tools in animal studies to show that the hippocampus also works to establish these rapidly forming cortical memories. The hippocampus helps to rapidly create immature memory engrams in the cortex."[69]

Simulations that occur with zero or immediate projection in time represent low-level concern. Animals that utilize this type of concern are responding to immediate threats. Take for example a pack of wolves bobbing and weaving through forestry in order to close in on their prey. They make adjustments to their formation and speed as a group in order to push their prey into unfavorable positions or circumstances. This demonstrates they have concern for a future outcome, the success of the pack, and wish to act in a way that will maximize their chances of reaching a positive outcome: capturing their prey to survive another day.

Given that humans are capable of simulating far off into the future, whereas this behavior has yet to be observed in any animal species, this type of behavior can be considered a form of dreaming (likely why we call it daydreaming). These simulations are under the full control of the daydreamer, which allows the learning process to take place without actually having had an experience. This type of behavior is what distinguishes the BCT from other animals and allows for the type of planning that is required to achieve TES. It is also the hallmark of the level of concern an organism possesses. Humans demonstrate concern for things that may happen in decades, centuries, or even billions of years in the future. This level of concern is the result of these brain networks evolving so much so that the ability to simulate has become intimately tied to what exists in long-term memory, as well as general creativity processes. This interlinking of processing modalities allows for the rapid growth of concern and intelligence.

The ultimate form of concern is that of the GCT that partakes in the commerce of the universe. This evolved BCT must ensure that it achieves ESS, which requires both simulation and creativity that far surpass what humans are capable of today. These requirements suggest that the GCT of the future will have enlarged left prefrontal cortexes and a much greater degree of interconnectivity between the brain centers. Perhaps their dreaming cycles will promote more rapid assimilation of information into memory. They may also possess the capability to alter their states of consciousness at will in a manner similar to how humans have evolved the ability to meditate in order to change their conscious states, though, more immediately. While various concerns drive the evolution of the BCT, the highest level of concern is that of a TCM wishing to become a GCT that has achieved ESS. As such, this is the greatest driving factor in the grandest of timelines for the *fungus*.

EOS

- Fossil record dated skulls to epochs
- Brain growth to body weight

Conclusion

When one studies all that has happened, as well as the ways in which it has happened, in our solar system and on planet Earth, one realizes that these processes are not necessarily unique to this star system. Anything that has occurred on Earth could have occurred elsewhere. Thus, the systems by which life has evolved on Earth are the same systems that are at work in other galaxies, and even within this galaxy in different star systems. Thus, the process by which a *fungus* has formed from abiotic factors in order to develop increasingly complex organisms on Earth is the same process that will occur somewhere else. Viewing what this *fungus* has iterated toward, informs what other fungi will iterate toward on their biospheres. This centers around generating centralized and specialized organs that house memories and process information on a bipedal frame. These organisms will promote creativity and problem-solving behaviors that will ultimately allow for the innovation required to produce technologies. The interaction between these organisms that possess increasing levels of concern and their technologies perpetuates an accelerated level of evolution. The ultimate driving force of this evolution becomes clear: the quest for achieving both TES and ESS in order to survive as a GCT. From the very first protocells existing as replicating RNA to the BCT, the spore of the *fungus*, every organism's contribution to evolution was to find a more efficient and successful way to survive. Carried forward, this extends to the cosmos and what lies beyond our galactic borders that can further our evolution and promote MCG.

4. TECHNOLOGY

Consciousness, Awareness, Concern, And Technology

Definitions (Sourced from Merriam-Webster)

Consciousness
The quality or state of being aware especially of something within oneself.
The state or fact of being conscious of an external object, state, or fact.

Awareness
The quality or state of being aware: knowledge and understanding that something is happening or exists.
Promoting a heightened awareness of the problem.

Cognitive
Of, relating to, being, or involving conscious intellectual activity (such as thinking, reasoning, or remembering).

Concern
A feeling of worry or care about a person or thing.
Note: Our prior use of the word concern does jibe with the definition here, as almost all cases of concern promote anticipatory behaviors.

Psychology
The sum total of all currently held beliefs (or bits of memory) that influences behavior/emotion.

Introduction

The aim of this chapter is to demonstrate that an organism is first conscious before it is aware. Through this awareness and the concern stemming from that awareness, technologies are developed through cognitive processes in anticipation of a need-based circumstance (concern) arising that is usually detrimental in nature. This cycle continues until the organism reaches a state where TES (Technological Evacuation Standards) is achieved and the organism is capable of becoming a TCM. As these technologies are developed, they branch out and recombine in a manner similar to that of a slime mold. This process includes conflict, cognition of various threats (both present and future), and planning. In this process, technology iterates toward an end goal of TES, so that all is not lost when the biosphere is brought to an end by various cosmological processes; usually the death of that *fungus*' host star.

Planning, an Anticipatory Behavior as Evidence of Intelligent Growth of the *Fungus*

TL;DR: A successful biosphere is the premature conscious-growth period of the *fungus*. It is only when the spore of that particular *fungus* has reached its TES and has successfully left its biosphere forever. This will have to be a state of ESS, by which it has then reached a matured conscious state, no longer needing any of its terrestrial resources to survive. It will take the entire final kingdom of technology to secure such a state of fungal growth.

ITM Planning is an adaptive behavior that arose from the continued conscious-growth of the *fungus*, which is evidence that increasing consciousness leads to awareness and that this awareness is expanding in a fungal manner. For example, the earliest SCCTs had very little consciousness in that they were only able to detect gradations in chemical concentrations in their environment and other internal measures of their state of health. With this limited level of consciousness, evolution occurred very slowly. It took billions of years for primitive animal life to evolve. However, specialized structures called organelles were forming in SCCTs during this time. These organ structures allowed an incremental but pivotal increase in the ability for these early forms of life to sense, or have consciousness, of the world around them. With this increased consciousness, greater awareness was soon to follow. With the arrival of MCCTs, centralized structures for memory formation and information processing allowed for behaviors that demonstrated greater levels of anticipation and awareness. For example, it wasn't until complex MCCTs began to explode in diversity that the first eye appeared in organisms.

"The first eyes appeared about 541 million years ago – at the very beginning of the Cambrian period when complex multicellular life really took off – in a group of now extinct animals called trilobites which looked a bit like large marine woodlice. Their eyes were compound, similar to those of modern insects."[70]

In evolutionary terms, it wasn't long after the first primitive eyes appeared that the more camera-like eye that we see in the world today would evolve. Thus, the exponential and fungal nature of evolution is beginning to demonstrate itself. What took billions of years to achieve is now being vastly improved upon in fractions of the amount of time.

"According to evolutionary biologists, it would have taken less than half a million years for the most rudimentary eye to evolve into a complex "camera" eye like ours."[71]

With the vastly increased sensory input of both primitive and more advanced eyes, these creatures were more conscious of their surroundings. Recall that increased consciousness relates to the increased sensory perception of an organism's surroundings. Therefore, with this increase in consciousness, further evolution in the memory and processing organelles of these early MCCTs was accelerated. These MCCTs are then increasing their awareness as they can now process what it means to be able to sense a whole host of things at much farther distances than their ancestors. Sensing light levels, predators in the distance, food sources near and far, as well as being able to have another metric by which to determine friend from foe merely gives these creatures greater ability to generate nuance from input.

This process continues throughout the fungal development of life on Earth, from the MCCTs up to the ACTs and QCTs. Each successive iteration of the *fungus* gained more consciousness and then, with time to experience this increased consciousness, increased awareness. This process continues in all organisms and results in adaptation and refinement of the structures that promote consciousness (i.e. the sensory organs), as well as the structure(s) that promote awareness (i.e. the organ(s) that process the conscious input). As this process has continued, the abilities of organisms to be anticipatory in nature have continued to increase.

An orangutan is more able to anticipate threats and changes in the environment than a fly. It is this natural progression of experience first and then developing understanding through continued sensory input, that allows for anticipatory behaviors to arise. Awareness increases by this process. This continued to happen all the way up until the appearance of the first hominids. As has been mentioned before, the changing environment forced hominids to stand further upright for longer periods of time, which had the effect of freeing the hands to be used for a multitude of other purposes. It is not to say that hominids couldn't use their hands for many tasks, as they could indeed do that; however, they were in fact freed up for more time that allowed the progression toward increased creativity and the creation of technologies such as tools. By the time the first BCTs showed up on Earth, the specialized processing organelle (the brain), had time to evolve to such a state of consciousness that the brain was processing multiple modalities of sensory input and had been doing so for billions of years. This exposure forces the evolving brain into further optimization and processing pathways that allow for ascribing meaning and order to the overwhelming amounts of data that are streaming into it. This is all done to enhance chances of survival, as well as to adapt to the environment, as was discussed prior. As these BCTs evolved further and further, they became more and more like the humans of today. This progression is easy to understand when we compare the abilities of modern-day or near modern-day humans to plan and anticipate when compared to other animals and organisms.

IPT Every great success achieved by humans can be traced back to a well-thought-out plan. Every failure is a result of the opposite: a lack of planning or poor planning. In humans, the more developed frontal cortex (as mentioned prior, specifically the left prefrontal cortex) in the brain allows for imagination, forethought, planning, and anticipation (concern) created by the hippocampus (memory). This allows the planning process that most other animals are incapable of, outside of seasonal preparations (hibernation) and hunting, foraging, and seeking shelter. While most humans are capable of this process, many humans are too occupied with daily concerns of survival in order to employ this ability in truly meaningful ways. Here, meaning should be thought of as promoting an understanding or awareness that drives the *fungus* forward. This is because these types of people are caught up in the societal machinery, where so many find themselves stuck between its grinding gears. This consists of bearing and raising offspring, working to prevent getting evicted, or having their utilities shut off, etc. The naturally gifted few are able to see beyond the minutiae and possess greater levels of concern (awareness) and civility, which results in greater levels of planning and success. These few are vital to the survival of the fungus as they push the envelope of what is possible and ensure that the fungus reaches TES.

BP Through millions of years of planning, due to concern and awareness, the fungus begins to develop a sense of its own impending demise as Earth will one day no longer be able to support life. This awareness stimulates an accelerated effort to produce technologies that will meet TES for future TCMs. This process is occurring on Earth and we are in the transition from being mostly unaware to being mostly aware of these limitations placed on the time we have to exist on Earth and subsequently in our solar system. This polarizing event is another catalyst in terms of the evolution of life on Earth. With all eyes focusing on the same goal, and all brains innovating toward that goal, humanity will progress at a startling rate in comparison to the chaotically aligned goals of the various governments, groups, and nations that exist on Earth today. If humanity is to have any chance at surviving Earth's end, the intelligence and awareness afforded to us by the evolution of the fungus must begin to be utilized so that we may become the spore of that fungus and live as TCMs via ESS in the cosmos.

BBP The same would be true for any civilization in any star system. The universe forces this result due to the nature and dynamics imposed by the three fundamental limitations of the universe. This pushes any sufficiently advanced conscious-growth toward creating technologies to meet TES and, ultimately, become a TCM. This universal nursery forces its spawn, formed on a multitude of different petri planets, toward greater levels of consciousness and awareness via a fungal process. That is to say that as life sprouts in one place or another within the universe, it sprawls outwards in a fungal manner, both by its increase in complexity and capability to generate technologies and by its potential to spread outwards from its birthplace. As mentioned in Chapter One, the fungal process of growth can also be seen in the formation of superclusters like Laniakea. It is easily seen that the guiding principles of the universe force both matter and life to take on similar characteristics at all scales. Each and every fungus that formed and evolved long enough to possess the ability to plan for its eventual biospheric evacuation is also likely planning for the possibility of meeting other life out in the universe. In doing so, it will have prepared itself in a manner that promotes its conscious-growth and awareness.

Conflict Stimulates Technology

TL;DR: The advancement of technology is often driven by the need for better tools in warfare, leading to innovations in both offensive and defensive capabilities. This historical trend highlights the constant struggle between survival and the development of technologies, ultimately shaping the course of civilization. The essay explores the idea that wars and conflicts may be essential for the evolution of technology, especially in preparing for Earth's survival. The argument suggests that the tools developed through conflicts become crucial for facing various challenges, leading to a symbiotic relationship between humanity and technology. Ultimately, technology is viewed as a replacement for Earth in the face of an inhospitable future.

ITM Historically, technology has almost always advanced more quickly due to humanity's need to develop technologies capable of dispatching their foes faster. The ability to achieve a faster kill has always won wars, not only for those on the offense but even more so for those on defense. From the first arrowhead to swords and knives, war has forced technology towards increasingly faster production and innovation cycles in order to avoid death. This has been achieved either by directly thwarting the enemy or by increasing the ability to survive through the use of technology that is defensive.

"In the second millennium BC iron tools and weapons began to appear (Wainwright, 1936; Coghlan, 1956, p. 61)...but the full potentialities of iron were not realized for many centuries."[72]

With these metals and alloys, both swords and shields, as well as other technologies, can be developed that possess both offensive and defensive capabilities. After iron tools were developed, centuries later, civilization began to develop alloys by compounding metals with other materials in order to increase their strength and durability. This led to lighter but stronger materials, as well as sharper blades. This increase in sharpness and durability, in effect, also reduced the time it took to defeat foes because fewer swings were necessary to pierce armor or behead an enemy.

"A vast range of alloys can be made but there have been few instances of the use of steel or iron alloyed with other elements than carbon, one being the metal made from the famous Noricum ores and another, the phosphorus-rich iron used by the Romans. However, not only can iron be alloyed, like copper, but when combined with carbon it can also be heat-treated to produce a vastly improved material, with a great range of properties, depending on the alloy content and heat treatment."[73]

In each instance, the need arose for technology in order to ensure survival. The primary impetus was not the consideration of the foe but only that of one's fellow comrades. Survival of allies was placed above any other considerations of what may become of the enemy, as would seemingly be common sense. However, after the haze settles and the fog of war is lifted, the horrors it brings inspire a different view of what could be done or how technologies could be utilized.

Technology First, Morality Second

In any war or conflict, there will be disagreements about the way in which the war is being carried out, either by the aggressors or the opposers, as well as how people may be being treated on either side of the fence. However, one thing is for sure: these grievances will not matter one bit if the side that holds these concerns is completely wiped out. This would suggest that morality is a luxury that comes only to the victor. The victor is usually the one that utilizes technology more effectively or generates new technologies that overwhelm their foes.

Once the conflict is over, considerations that relate to how the war was carried out or how people were treated can be entertained. This would suggest that morality comes secondary to the growth of technology. As much as humanity would like to believe that it is a moral species, the *fungus* cares not one bit one way or the other. The goal of the *fungus* is to ensure that TES is achieved by whatever means necessary. In this sense, what dies and what lives are of little consequence in the greater sense, as there would be no one left to be concerned about such things should the *fungus* fail and all life on Earth be wiped out as it overheats.

In essence, this is a formula that must play out in the *fungus*. This process repeats at every level of the *fungus*, as well. The Monkeydactyl was never going to be the creature that became a TCM; however, the *fungus* was utilizing what had been learned by that iteration to bring the *fungus* forward. If morality were to be considered, was that right to "use" the Monkeydactyl in this way? The point is that to the *fungus*, it is irrelevant. All that matters is the end goal of producing a spore and that spore will be thrust away from this birthing ground into the cosmos as a TCM. Any other considerations come secondarily. That is not to say that morality isn't a useful tool that may assist in the development of future technologies or societal structuring. However, it does necessitate that these morals cannot be a primary consideration as there is the potential for failure if the primary concern is not survival.

IPT Fast forward to World War I and technological advances occurred not only in offensive technologies but also in life-saving medical advances. This occurred due to the need for saving more and more lives due to the fact that the speed at which lives were being taken had increased drastically. The advent of medical vaccines, mobile X-Rays, antiseptics, the splint, and the ability to treat burns were pioneered during this time.

"The weaponry was far more effective - 15% of all combatants in the British armed forces were killed and that was unprecedented...On the other hand, of those that got injured but didn't die, far fewer than before died of their injuries and that has to be down to well organized medical management...With hundreds of thousands of injured soldiers returning home, World War One also led to a new emphasis on rehabilitation and continuing care. New techniques in facial surgery and burns were developed - and there were huge advances in prosthetic limb technology - to meet the needs of hundreds of thousands of amputees."[74]

During World War II, many advances that revolutionized the way we travel and sense the world around us were made. Radar, which makes the detection of objects at vast distances possible, allowed for early warning of incoming attacks. The advent of the jet engine also occurred during this time. The ME-262, which was the world's first operational jet fighter, changed aviation forever. It allowed for speeds that exceeded the prior propeller-driven planes as well as paved the way for the modern jet airliner. After World War II ended, the world was amazed at the V2 rocket used to bomb England because it was such an advancement in aerospace and rocketry. It was also the first object launched into lower outer space. The scientist responsible for its success was Wernher Von Braun, who eventually surrendered and was successfully captured by the United States. When asked how he came up with such an unbelievable machine, Von Braun stated that he had, in fact, borrowed the idea from the American physicist Dr. Robert Goddard.

"Dr. Robert Hutchings Goddard (1882-1945) is considered the father of modern rocket propulsion. A physicist of great insight, Goddard also had a unique genius for invention. It is in memory of this brilliant scientist that NASA's Goddard Space Flight Center in Greenbelt, Maryland, was established on May 1, 1959."[75]

With the ability to attack targets at a much greater range, the potential for mass death and destruction, or at the very least, harm is greatly increased. Therefore, this improvement in offensive technology that allows for attacks to occur at such a distance that an opposing force would never see their enemy coming, would logically result in the necessity for advances in the treatment of injuries and ailments. Therefore, offensive technologies like the rocket often inspire the advent of better medicines and surgical procedures. This demonstrates that the development of technologies that kill with greater speed and efficiency also stimulates medical advancement that helps ensure humanity's survival. As we progress toward the modern day, technologies like duck tape (now called duct tape), anesthesia, organ transplant, chemotherapy, plastics, polyester, computers, and many other amazing and transformative technologies have become an integral part of daily life. These technologies piggybacked off into the technologies of prior generations of which war was a major catalyst.

"Once the principles of military surgery were relearned and applied to modern battlefield medicine, instances of death, deformity and loss of limb were reduced to levels previously unattainable."[76]

Along with the obvious threats that come with the advent of technologies that can kill with great speed and efficiency and do so over extremely large distances, it is the awareness that this circumstance stimulates that is also a necessary set of ingredients for bringing about technological advancement. Oftentimes, these circumstances, such as was the case when the first atomic bombs were being developed, bring about a level of awareness in the *fungus* that generates a significant amount of concern. This concern is primarily focused on the amount of human life, and possibly other life existing in surrounding habitats, that would be lost if a specific technology were to be utilized or not. The concern impaired by this threat generates a response that demonstrates a greater love for human life, however paradoxical it may seem. To kill a large group of people to end a conflict or war versus prolonging it is demonstrating an overall level of concern that is greater than keeping the status quo. Wars and conflicts often promote this type of concern and give a glimpse into the fact that the love of life is stronger than the desire to kill. A great example of this is what happened in World War II.

"Truman stated that his decision to drop the bomb was purely military. A Normandy-type amphibious landing would have cost an estimated million casualties. Truman believed that the bombs saved Japanese lives as well. Prolonging the war was not an option for the President. Over 3,500 Japanese kamikaze raids had already wrought great destruction and loss of American lives."[77]

BP Is it possible that all these wars and conflicts are a necessary component of the *fungus* evolving and preparing for TES? Would we have developed each of these technologies in the same manner if Earth remained peaceful throughout history? We don't have any long-standing examples throughout human history of whether or not these conflicts are more beneficial to the stimulation of technology than a comparable period of time without any form of conflict. However, up until modern times, war and conflict have been the leading forces behind technological advancement that ultimately improves the quality of life for most people on Earth. When Earth's habitable days are drawing closer to an end, technology will be all we have left in order to save ourselves. All of the wars and conflicts that we have endured throughout the ages will be given new meaning when it is realized that they inspired the technology that allows us to survive. This is a good example of the *fungus* self-stimulating in order to generate technology for TES.

The tools and technologies that have been developed through conflict and war will likely always end up becoming a necessity in order to weather the storm of an increasingly harsh and hostile existence, which includes natural disasters, changing climate, and manmade conflict alike. Much as animals also rely on other kingdoms of life, such as plants and eubacteria, humans rely on technology. This reliance has only increased with time, akin to a symbiosis evolving between organisms in nature. This usually occurs through many generations and takes time to reach stability. Ultimately, all life will depend on technology for its survival as technology is the construct that will end up replacing all of that which our planet provides. Technology will end up being the replacement for Earth itself when Earth is no longer habitable or destroyed. As mentioned before, this new Seventh Kingdom (the Technium) operates in a form of symbiosis with humanity. It is because of this that the Technium is a part of the *fungus* that was inevitable and is Evidence of System (EOS) to the human condition of absolute (HCOA).

BBP What purpose would the creation of any life really serve if it was not always doomed to be destroyed by the very biosphere that fostered its development? Each and every conscious-growth that should evolve in any biosphere will ultimately develop some form of technology in order to survive the threats of that biosphere and become a TCM. The knowledge that this conscious-growth's demise is inevitable is the driving force behind MCG and is inescapable as it is literally baked (temperature funnel) into its existence. As such, war technology must be viewed as part of the *fungus* wherever a fungal colony (conscious-growth) should form.

The Symbiotic Relationship between Consciousness, Awareness, Concern, and Technology

TL;DR: The essay explores the symbiotic relationship between consciousness, awareness, concern, and technology. It suggests that as awareness evolves in conscious entities, there is a drive to fill existential voids, leading to the simultaneous development of technology. The discussion includes how awareness in organisms, limited by their complexity, influences their concerns. The analogy of fungus as a collective consciousness is presented, and the early biological adaptations, like fungi producing penicillin, are discussed as early forms of technology in response to threats faced by life on Earth.

When awareness increases through evolution, a conscious entity will attempt to fill the VOP (recall that the VOP was introduced in Chapter One as a void in the identity of a consciously aware entity or a collective consciousness) and, in doing so, force technology to develop with it, as one cannot advance without the other. First, there is a growing awareness that comes as a result of being conscious that all living things, including humans, are subject to a limited food chain, which keeps a consistent threat of starvation. If any living thing has made choices to survive these threats, it is due to being made aware of the second fundamental limitation of the universe: inevitable death.

Ultimately, it is the successive conscious experience of various threats and the adaptation to these experiences through awareness that promotes concern in organisms. This awareness is limited in each animal, plant and fungi, based on the complexity of their structure and/or brain – their level of consciousness.

It is hard to think of the fungus as being conscious or a form of conscious development. However, if you account for the billions of growth network contributions combined, developing under different conditions as temperatures change, it represents a developing collective consciousness – which is the fungus. Conscious awareness develops and spreads in a branching and merging structure, similar to the behavior of slime molds as they spread out from some central point and move toward sources of food. However, it is an uncontrollable process once it catches hold.

ITM As early life on Earth began to develop, it instantly faced various hurdles to overcome. Some organisms could not move; some could not utilize sunlight, and others were incapable of digesting/ingesting food. Each one of these hurdles that was to be overcome required a biological adaptation at this point in time. Life (the fungus) was not complex enough to create physical structures and technology externally. One such example of this sort of thing happening in the bacterial/fungal world was the discovery of fungi's adaptation to bacterial threat through the production of penicillin. Penicillin could be thought of as an "artificial" creation by fungi in order to defend against other invasive bacteria. In this sense, the earliest technologies that were developed on Earth occurred during the beginnings of life on Earth.

"The antibacterial effect of penicillin was discovered by Alexander Fleming in 1929. He noted that a fungal colony had grown as a contaminant on an agar plate streaked with the bacterium Staphylococcus aureus, and that the bacterial colonies around the fungus were transparent, because their cells were lysing. Fleming had devoted much of his career to finding methods for treating wound infections, and immediately recognized the importance of a fungal metabolite that might be used to control bacteria. The substance was named penicillin, because the fungal contaminant was identified as Penicillium notatum. Fleming found that it was effective against many Gram positive bacteria in laboratory conditions, and he even used locally applied, crude preparations of this substance, from culture filtrates, to control eye infections."[78]

Fungi of the very distant past would have had other strains of bacteria that they were in close proximity with and ultimately would have to defend against. These early fungi would have been conscious enough to determine that other entities were encroaching upon their surroundings and may have possessed an actual awareness which would have had to adapt in a very similar way in order to stay competitive in their niche environments. Thus, whatever products these fungi were capable of forming to stave off bacteria growing nearby would have been an artificial product that had not existed prior to its generation at that time. Thus, early technology was likely being developed on Earth as early as life itself began to evolve. This demonstrates that the consciousness of fungi allowed for an awareness of what threats to its existence would mean in the long run. Thus, an adaptive and anticipatory behavior displayed by the fungi was the generation of defense biotic factors in order to survive.

Fungi and bacteria are not the only examples of organisms utilizing some form of technology in order to survive or improve their chances of survival that stems from the awareness that harsh competition exists in a world filled with competing organisms consuming the same resources. After multicellular life began thriving and the first ACTs began to show up, some of these organisms began to develop hard external surfaces that acted as protection from the outside world. When we see snails, turtles, and other shelled creatures, they are the descendant of these ACTs that first developed an outer-surface technology in order to protect themselves.

This ability to generate technology as part of the organism exists throughout the history of the *fungus*. However, it is not until animals like birds and QCTs emerge that this capability extends outside of the organism and into interaction with the environment in any appreciable way. By the time the first hominids began to evolve, these BCTs were beginning to stand upright more and more. With their opposable thumbs and the extra dexterity in their hands due to being able to stand upright for more than a few seconds, they were able to create tools, albeit primitive ones. Sticks were likely the first tools to be used to extract insects from mounds. Not long thereafter, archaeological evidence demonstrates that primitive stone tools, spears, and knives were being crafted by early hominids. Further evolution would bring about a fully upright species with unrivaled crafting capabilities, as well as an intellect that could allow for the continual refinement of their skills in creating tools and weapons as they became further aware of the necessity to defend themselves and procure food. After these upright hominids evolved further, the first humans appeared on Earth.

Early humans were no exception to this pattern as they began to develop tools and weapons just as early hominids did. This ultimately led to larger groups of humans being able to inhabit any particular area due to the fact that they had easier access to food. With increasing population size, civilization began to advance as well. Once the use of these tools became widespread, civilization advanced at a faster rate. However, these technologies are dangerous and can pose a threat to other humans as well. When early humans began using these technologies against each other in their territorial disputes, it was concern over survival that developed defensive technologies that could ensure continual survival. Through the continual advancement of technology, humans have been able to survive both the threats forced upon us by the conditions of our planet, as well as the threats posed by other humans. With each successive technological advancement, civility increased as humans began to fear the consequences that come from that technology being used against them.

When early humans began to inscribe images on the walls of caves, this represented a massive jump in conscious-growth. It could be said that this was the moment that humans were truly self "aware" of their existence. They knew that they existed and pontificated (self-reflecting) things they may not have been able to convey with their primitive forms of communication, as no such formalized way of storing the information existed. However, they knew that if others were to look upon these images, they would be able to relate in some way. This represents a major leap in the consciousness of humans and the *fungus* as a whole, as humans were now processing more inputs in their advanced brains than other animals. This is a major threshold to cross on the path toward TES as it is a requisite step in order to become aware of one's place in the cosmos.

Similarly, certain events occur that are entirely transformative and increase the awareness of humanity drastically.

An example of such a transformative shift taking place would be the increase in awareness that was a result of the invention of the lens. The lens was not initially used for looking skyward; however, not long after its appearance, humans began to use it to peer into the heavens. The realization that we exist on one of many planets in our system, and that there are many such star systems out there, broadened the scope of human understanding quite significantly. While it took many years of debate, inquiry, and discovery in order to conclude that there are other star systems and galaxies in the universe, it was all sparked by merely being able to look outward into the heavens. This period of time, the Optical Revolution, is one of the defining moments in the change in awareness of the *fungus*. The reason this is such a profound moment is that it gives the *fungus* the ability for when Homo sapiens finally evolve, allowing humanity an understanding of the galactic positioning of our *fungus* in the universe, and the dynamics of its solar system, as well as informs us that there is vastly more out in the universe than what we can see with our naked eye. This information allows the collective consciousness of the *fungus* to begin to toy with the idea that life "a *fungus*" elsewhere may also exist. Given that the number of other star systems is so great due to the number of other galaxies, the idea becomes much more plausible than what earlier religious beliefs would tend to portray. This increase in awareness, and others like it, allow for new concerns to be brought up and addressed in order to push the fungus forward.

Ultimately, the key takeaway here is that technology initially progresses slowly but picks up its pace in an exponential manner. This is due to the fact that organisms take significant time to evolve the requisite sensory organs that generate consciousness. However, with each incremental increase in consciousness, the organism becomes more aware. This is akin to the number of tendrils/arms of an expanding and merging slime mold. Once again, we see that the universe is pushing things forward in a fungal manner. In the same way that you cannot control the spread of a rumor, you cannot control the spread of technological growth.

IPT Continue this line of reasoning in modern times, where nuclear weapons are used to deter entire countries from engaging in war with one another, and the pattern becomes clear: the ability to become more conscious throughout the evolution of the *fungus* has led to an increasing awareness over many eons that promotes a concern for continued survival, which ultimately stimulates technological advancement. As mentioned before, and while sounding paradoxical, the development of increasingly malicious and deadly technologies that speed up the rate at which deadly force can be used promotes peaceful behavior. This occurs because the fear of its use outweighs the desire to protest, which results in more docile behavior. The group or civilization with the technology that can kill the quickest is always the one that stands above the rest. However, this forces those with lesser technologies to strive towards ousting those in superior positions. This cycle continues, which leads to increasingly civil behavior, and, subsequently, the advancement of civilization over long periods of time. That is not to say that, while this may be true, those on the side that end up employing these types of deadly technologies will not be opposed to their use. One way in which humans try to communicate complicated feelings about tenuous moments throughout history is through the arts. The arts are able to speak to a wide variety of audiences without necessarily having to be too technical. By speaking to people in a way that resonates with them on a more day-to-day level, films are able to make large groups of individuals aware of something they may not have been before. One such example is Stanley Kubrick's *Dr. Strangelove* or: *How I Learned to Stop Worrying and Love the Bomb.* This film attempted to make people aware of the potential dangers of these extremely powerful technologies being placed in the hands of people who might not be the best individuals to be left in control of it, while casting such a topic in a humorous light. Praised as arguably the greatest political satire ever made, Dr. Strangelove revolves around an unhinged U.S. general single handedly launching a nuclear attack on the Soviet Union, effectively kickstarting World War III.[79]

While this film's main premise was initially rebuked by the government as being absurd and totally impossible, humanity would later find out that such mechanisms that allowed for the calamity in the film to take place actually did exist and operated at the governmental level in much the same manner as depicted in the film. It is very possible that Kubrick's film, while initially frowned upon by officials, may have had an impact that kept humanity from entering into a devastating Third World War.

"Various defense intellectuals and government officials dismissed the film as 'impossible on a dozen counts'...However, we now know that top U.S. military commanders indeed had presidentially authorized instructions providing them with the authority to launch nuclear weapons. As declassified U.S. documents show, such pre-delegation existed beginning in 1956 when then U.S. President Dwight D. Eisenhower authorized U.S. air defenses to use nuclear weapons to defend against

Soviet bomber forces in the event of an attack.'[80]

Through the use of film and other media-based technologies, humans now possess the capability to promote greater awareness, which in turn promotes more technology and, hopefully, more responsible use of that technology. We have not utilized a nuclear weapon since its initial use in Japan, as doing so now would almost surely result in the near-total destruction of humanity. There are so many nuclear weapons that are housed in silos around the world today that if even a fraction of them were to be launched, the world could be destroyed many times over.

The modern-day human now has more than just the ability to watch movies in their home or out in theaters; they also are able to play games with or without others online, as well as interact through a variety of streaming platforms where content is created. Each of these developments, while seemingly unimportant for the species in terms of getting us off the Earth, iterates humanity one step closer to the necessary technologies that will allow us to achieve TES. As an example, streaming and content delivery requires so much data to be trafficked through the internet that advancements in the transmission and compression of data in a lossless manner have evolved significantly. The influence that these seemingly trivial developments in leisure have had on the capabilities of our civilization is anything but trivial. It will be essential to transmit extremely large amounts of data over very long distances with little-to-no loss in fractions of seconds for humanity to survive as a TCM out in the cosmos. This is just one example among many where technology is driven by the developments that occur in the day-to-day lives of humans engaging with technology. At one point or another, the focus of the average individual in the collective will no longer be set on surviving the next attack or the next famine but on surviving the end of our entire biosphere. It is at this stage that conscious awareness shifts from territorial/resource concerns towards sustaining life through self-sufficient technologies–aka ESS. At this level of conscious awareness, where each and every human understands that mankind's time on Earth is limited, humans will all be working together in a sort of mind-hive in order to ensure the survival of the species. There will be no other goal that is more prescient as the survival of humanity will depend on the combined effort of every single human.

Without technology and the increased awareness it brings, there is limited intelligent growth. This growth must happen slowly, though, so that what is added to the collective consciousness is both truly understood and that the subsequent technologies developed are stable and reliable. Modern examples where this is made apparent are aeronautic and aerospace engineering, data storage and retrieval, and the vast multitudes of machines that operate in industrial settings to produce our foods and goods. Looking forward, humans must place immense amounts of trust in the technology that will be developed to support our biological functions once we must permanently evacuate the Earth. Since technology improves when conscious awareness increases, the two move in tandem. Due to the fact that technology grows and spreads out in various arms and branches, its progression is also a fungal process. Since technology must develop slowly, it is afforded ample time to be assimilated into society and is subject to rigorous testing and refinement. This ensures that society can move forward incrementally through the technology it develops. From the frame of reference of an individual, the perception of experiencing life may seem somewhat static. It would appear as if not much is changing from day to day. However, the universe makes sure there is very little leisure time that does not move technology forward as the Earth gradually expires. Refer to the figure that contains the graph of technological growth in the last 200 years.

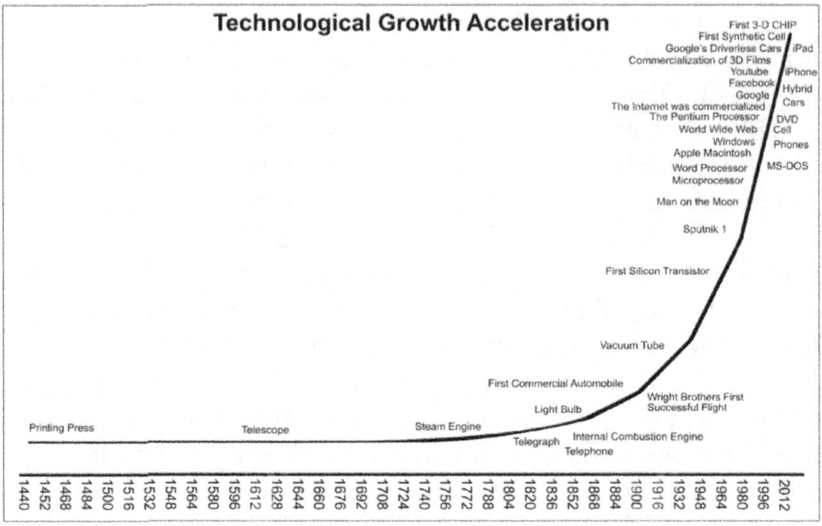

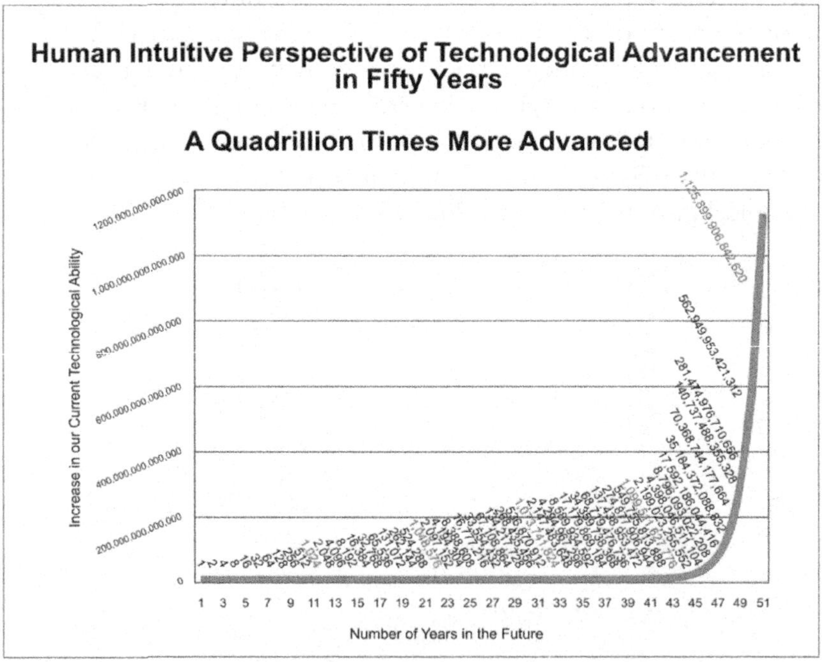

Figure NN78 Accelerated Technology

In the top portion of the figure, the graph illustrates the various technologies that have been developed from the printing press up until the development of the first 3D chip. Each technology is, on average, many times more complex than the prior. The growth of this technology is explosive and leads to the rapid development of other technologies. As mentioned before, this also pulls conscious awareness along with it. The lower graph in the figure represents an exponential growth curve, namely 2x. Here, given an initial value of 1 for technological complexity, the value doubles each time. This is in line with Moore's Law, which states that "the number of transistors on a microchip doubles every two years." Comparing the two demonstrates how technological growth is an exponential process and, in the invention of new technologies spanning multiple different fields, the process looks fungal in nature as the "terrain" of different possibilities in technology is explored. This causes accelerated development of conscious awareness as a result. In most cases, it is true that without a preceding technological advancement, the subsequent advancements would not be possible (they usually piggyback off one another or recombine like the branching networks of a slime mold reaching out to find a path toward TES).

BP Conscious awareness has been continuously developing, from the very first single-celled organisms up until modern times as one entity that grows and spreads in a fungal manner. It is a gradual process that, by the end (spore stage), produces technology in a universe that operates through HTO in order to promote survival after Earth's end. As humans continually develop newer and more sophisticated technologies, a new kingdom of life is in the making. This eighth kingdom, or Technium, houses all technology that has been or will be developed by man. However, in the not-so-distant future, computers will design and manufacture themselves. This is already occurring in some very specific and limited applications. When A.I. becomes self-aware, technology will truly have reached a sentient status and thus be a fully-fledged kingdom of life. This kingdom, after some maturation, is a natural and necessary extension of biological life trying to prolong its existence and adapt to the harsh conditions of surviving as an extraterrestrial.

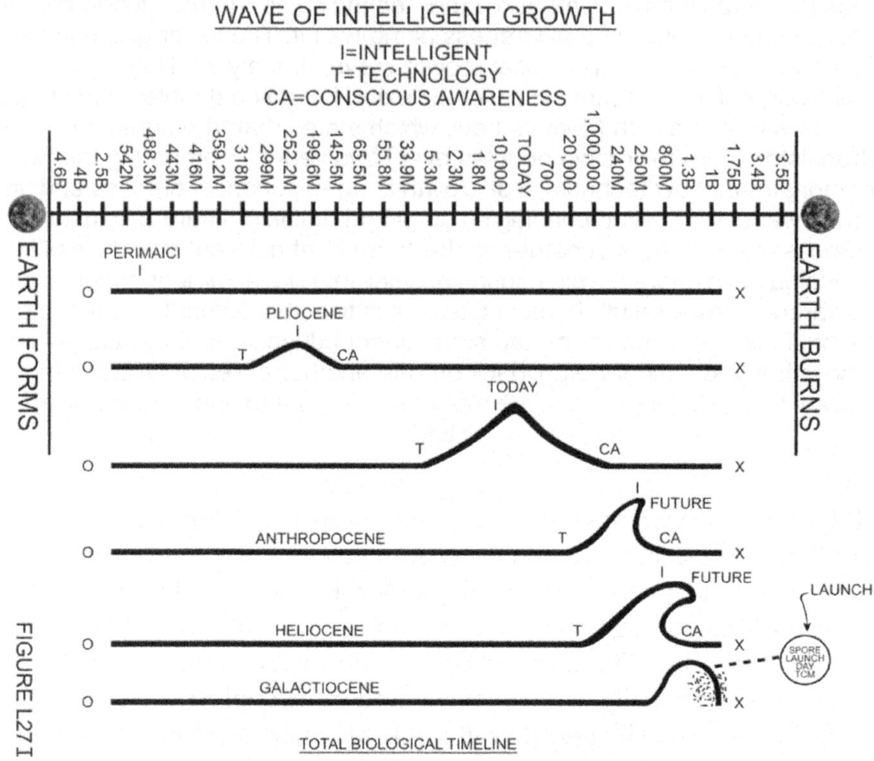

Figure L 27I (Wave of Intelligent Growth)

This figure demonstrates how intelligence grows like a wave through the many epochs of Fungal development, even with launching into outer space by the end of its biological terrestrial and premature formation.

BBP Starting from the first replicating RNA structures and single-celled organisms, and ending with the humans that have achieved TES, one conscious awareness growth encapsulates this entire organic development. Just as this process will happen on Earth, it has either happened elsewhere in the cosmos, is happening, or will happen. The universe is the breeding ground for consciousness and the awareness that it affords. There are countless planets in different configurations of size, orbital distance, composition, and density. Some combinations promote life while others do not. Given the vast scale in both size and time, other fungal growths are sure to exist. As these fungal growths go on to achieve their respective TES, they will come into contact with other fungal growths should they not be the only life in the universe at that time. This advances the conscious awareness of that *fungus* further, through interacting with and evolving from changes imposed by these encounters. Ultimately, the exchange of technologies, experiences, and culture drives increased conscious awareness until MCG is reached. All consciousness strives toward this point for the UGCH.

Likening Technological Advancement to the Behavior of Slime Mold

TL;DR: The essay draws parallels between technological advancement and the behavior of slime mold. It explores how seemingly unrelated inventions and innovations from the past can converge to create transformative technologies. Using examples from the evolution of computing, it demonstrates how technologies, initially developed for different purposes, can synergize to form new, more advanced inventions. The evolution from punch cards to magnetic tape drives, and eventually personal computers, illustrates this interconnected progression in technology and information transfer much resembles the same way of slime mold.

In order to understand how the evolution of technology is likened to that of the growth of a slime mold, it is instructive to view how seemingly unrelated discoveries, inventions, and innovations from the past may come together at a later point in time and be utilized in order to create something transformative, oftentimes superseding each of the original elements. Thus their combination created some form of synergy. The fungus utilizes this behavior, so it is no surprise that the Technium would as well. In general, technologies are not only ever improved by becoming further specializations of something prior but are often used in combination in order to create something novel.

IPT The personal computers we use today are vastly more powerful than the first computers that began being used in the early and mid-to-late 1900s, however, few are aware that this technology was employed by the United States during the 1890s census. The most basic computers at that time utilized punch cards in order to count, tabulate, and record information. Punched cards date back to the late eighteenth and early nineteenth centuries when they were used to "program" cloth-making machinery and looms. In the 1880s and 1890s, Herman Hollerith used them with his tabulators—a core product of what would eventually become IBM.

However, these computers, specifically something that can perform tabulation, are not the only human invention capable of such a task. The abacus was one of the first known devices utilized by humans to aid in complex computations. The first known use of the abacus was by the Babylonians. However, more common forms known today are the ones utilized by the Chinese.

"The abacus, an ancient calculating device…originated in Babylon around 2400 B.C.E. as a 'counting box'. It was the world's first calculator and contemporary versions are still in use today."[81]

What can be gleaned is that these tools were first developed due to the need for an aide to assist in complex computations. Later, a similar need to tabulate and store pieces of information required a similar but distinct innovation. The abacus was limited to numerical computation/calculation. However, the punch card technology utilized in the late 1800s was capable of storing information as well. As these punch cards became too inefficient for use in information-dense settings, magnetic tape drives came to replace them. However, magnetic tape drives utilized for the storage of data are strikingly similar to the phonautograph, which was the first device to record the sound of a human voice onto a semi-permanent medium.

"The question of which sound was the first ever to be recorded seems to have a pretty straightforward answer. It was captured in Paris by Édouard-Léon Scott de Martinville in the late 1850s." 82

The technology that allowed sound to be recorded on a medium would influence future technology to allow data storage on a very similar medium. This is an example of separate technologies used for seemingly unrelated tasks being utilized in a combined fashion to create a new technology that supersedes the original invention/innovation's capabilities.

"In 1877 Oberlin Smith learned of the magnetic sound recording device (Phonograph) by Thomas Edison and decided to improve the Edison phonograph without initially getting away from a magnetic sound recording…The magnetic tape was not used to record data until 1950, when the UNISERVO I, the first digital tape recorder, was developed in Eckert-Mauchly Co. and was implemented in 1951 UNIVAC I computer architecture."[83]

This innovation revolutionized the world as data could now be stored and read in a much faster manner than had ever been achieved. However, this ability to store and analyze data would pave the way for the computers of today. While the UNIVAC I computer took rooms and rooms worth of space, the first Apple personal computers could sit on a desktop. With the increased portability of the personal computer, more and more individuals gained access to the capabilities of computers, which improves overall awareness by allowing for the improved dissemination of information.

Most everyone is familiar with a processor but is unfamiliar with the fact that computers did not always utilize these microprocessors. The advent of semiconductors led to the use of microprocessors and allowed bringing down the size of computers. In the early 2000s, computers with power much greater than that of the desktops designed in the late 1900s could fit into a pants pocket. With the improvement afforded by these advancements, human understanding of physics, chemistry, medicine, and even our own behavior has rapidly advanced. With respect to physics and chemistry specifically, these technological advances, as well as our understanding of the laws of the universe themselves, have paved the way for an entirely different shift in the paradigm of computation. Quantum computers are the result of all of what has been learned through physics and the advances brought about by classical computers over the last 100-200 years. Quantum computers utilize the interesting behavior of atoms, such as their spin, as well as the relationship between these properties when these atoms are "entangled," in order to run computations that would take classical computers thousands or hundreds of thousands of years in some cases. This is because quantum computers do not store information in "bits", as classical computers do. Bits are used to represent data in the form of 1s and 0s (1 for on and 0 for off). In quantum computers, analyzing the state of an entangled particle can return a 1, a 0, or a seemingly paradoxical combination (superposition) of 1 and 0 at the same time! The uses of quantum computation are seemingly endless and humanity is only beginning to truly unlock the potential for what these types of computers can do.

Another example of the way in which technology's growth can be likened to that of a *fungus* would be the invention of the plow, which revolutionized agriculture from that point forward. The first plows were invented approximately 6,000 years ago and consisted of basic sticks that would drag through the soil as the plow was pulled by an ox. The Egyptians were responsible for the first plows. However, other groups later improved upon the designs by adding wheels or improving the mechanism which displaced the soil. Regardless, this technological advancement allowed human cities to grow in size and improved the average health of members within these cities, as food became more readily available to evenly distribute. In so doing, civilizations began to thrive, giving more time to innovate and iterate on other technologies.

Fast forward to more modern irrigation and farming techniques, as well as sewer systems that were needed to displace the extra waste generated by increasingly large societies, and mankind's ability to manipulate its surroundings, transport goods and waste, and study without the threat of starvation. This opened up humanity to the world of science in general. These advancements all come back together when one sees how science has now begun enhancing agriculture. Basically, the plow made it possible for technologies that make it totally unnecessary for humanity to continue to use the plow. In modern times, machines that can drive themselves in the form of automated agriculture take the place of any human workers in some farms. CRISPR and other technologies related to gene editing can allow the generation of crops that grow in conditions that may not require plowing at all. By understanding the enzymatic processes that occur in fungi and bacteria, scientists have even been able to create food and other materials without the need for farming at all!

"During the 1970s space program, NASA scientists explored a way to feed astronauts on long space journeys by transforming elements in the air the astronauts breathed into protein. This process was shelved and forgotten back then, only to be rediscovered by Dr. Lisa Dyson and Dr. John Reed and perfected by the Air Protein team. Combined with modern food-making technology, we're delivering the most sustainable meat alternative on earth."[84]

With this technology, humanity takes itself one step closer to ESS as taking microbes along into space, or any other location for that matter, essentially requires no space at all. Minimal starter cultures can be brought along and then expanded as necessary through replication after ingesting the raw materials necessary in order to divide and continue producing protein as an output.

This behavior that technology possesses, akin to that of the slime mold, has not gone entirely unnoticed. James Burke, a famous science journalist, discusses these seemingly unrelated technological advances and how important they are in coming together to inspire the advent of technologies that come at later points in time, some of which are extremely far removed from one another.

"The PBS series is not your straight line, period-by-period examination of historical event-making. Rather, 'Connections,' written and narrated by James Burke, a BBC science journalist, is a circuitous, associative series of intellectual leaps that chart a pathway through the history of ideas, scientific inquiry and accidental discovery."[85]

James Burke is also able to demonstrate that technology advances in a way that first branches outward into multiple different avenues but comes back together far down the line to produce something truly transformative. This idea about technology isn't new; however, it is the way in which this behavior is classified and likened to a slime mold that is. This model, as is discussed many times throughout this book, aims to demonstrate that it is not only technology that advances this way but all of life in the universe–even the universe itself.

"...James Burke examines the ideas, inventions, and coincidences that have culminated in the major technological advances of today. He untangles the pattern of interconnecting events, the accidents of time, circumstance, and place that gave rise to major inventions of the world..."[86]

BP The rapid growth of technological innovation resembles fungal growth because of the countless micro-inventions along the path to each major innovation spreading out like the tendrils of a slime mold and later recombining in order to produce innovation that has the potential to transform the world. Actual fungus grows toward sources of nutrients. However, the *fungal* behavior of technology forces the Technium in the general direction of technological advancement that supports the development of a spore: the TCM. This TCM works in conjunction with the Technium it gave birth to in order to achieve ESS as a GCT. Ideally, working in tandem with these entities of the Technium, humanity will stretch further and further into the cosmos, promoting conscious-growth.

BBP To that end, there will eventually be a time when verbal communication ceases to be the primary modality of communication. The continued evolution of our species, and the requirements for both faster and more information-dense transfer of information, will shape a whole new system of communication in humans. This new form of communication will likely be one that shifts to direct communication between the minds of two individuals, as in telepathy. This will happen because the pace at which the brain processes information will begin to overwhelm the pace at which information can be delivered verbally. As an example of this, YouTube and other media providers currently allow the option to view content at up to two times the original pace. This demonstrates that there is a demand for increasing the pace at which information is delivered. The relatively slow pace of verbal communication is hindering both our response rates as well as forcing our brains to comprehend and ideate at speeds much lower than it is actually capable. This will eventually become a danger to the end goal of reaching TES for ESS as a TCM.

As with all other processes, this is occurring in other parts of the universe as well. Whether a species is more or less advanced than we currently are, they are moving forward and evolving. Perhaps they are at the stage of just beginning to write things down on the walls of various rock formations on their home worlds, or perhaps they are communicating with minds at vast distances throughout the cosmos telepathically. In either case, these societies are moving toward an increased form of consciousness and awareness through improvements in both technology and their own biology. As these societies progress, they inch closer to MCG. Eventually, it is possible that interspecies communication that transcends any form of verbal structure or written alphabet may become the norm. This is the *fungus* sprawling out and connecting with other networks, like the symbiotic relations in the mycorrhizal networks that exist between plants and fungi. Networks of fungi coming from different reaches of the cosmos could potentially begin sharing information and communicating in a mutually beneficial arrangement, as well. This occurrence surely promotes conscious-growth and advancement toward MCG. When the last beings to be spawned in the expanding universe have all but evolved to this form, MCG is all but achieved and the universe is ripe for the UGCH.

Communication & the Rate of Information Processing for Survival

Communication and the rate at which ideas, thoughts, and concepts can be delivered are essential for the proliferation of any species. However, any piece of information being communicated to another member of the *fungus* can only be understood if the conscious awareness of the other member is sufficient to be cognizant of the value of that information at the time. This demonstrates that the *fungus* develops as a progressive growth. This means that concepts by the whole can only be understood by the *fungus* once sufficient time and conscious-growth are allowed. Any time period before this moment will result in concepts being presented to a collective consciousness that is incapable of understanding them due to insufficient levels of conscious-growth. This is the behavior of all fungal growths. For instance, explaining to an indigenous tribe how to build or rebuild a car will be incomprehensible information. Thus, this section describes improving communication rates of ideas only in the sense that the information being communicated is of the appropriate level for that conscious-growth at the time.

Thus, enhancing and improving the rate of transfer, as well as information density, in communication is a part of evolution in organisms that are near the forefront of the *fungus*. This has been one of the leading forces behind the success of all microorganisms, plants, and animal life. The faster an organism can gain knowledge of an event and relay it to others within a community or society, the more quickly a response can be mounted.

ITM In the earliest organisms, communication as we know it today was not present, nor was it even possible. The earliest microbes were governed by sensing chemical gradients, or changes in the levels of chemicals in one direction or another. It is these gradients that would inform these primitive organisms where threats, food sources, or safer conditions may exist. This process, called chemotaxis, is pivotal in the ability of these single-celled organisms to move about their environments and survive.

"Chemotaxis in microbiology refers to the migration of cells toward attractant chemicals or away from repellents. Virtually, every motile organism exhibits some type of chemotaxis. The chemotaxis responses of eukaryotic microorganisms proceed by mechanisms that are shared by all cells in the eukaryotic kingdom and generally involve regulation of microtubule- and/or microfilament-based cytoskeletal elements."[87]

While this process describes movement in the direction of an attracting chemical or away from a repellent one, the idea of utilizing chemicals for some form of messaging is also used within multicellular organisms as well. Chemical messengers inside cells are also utilized in order to carry information about the state of affairs and workings of the different parts of the organism. In essence, early communication was chemical-based, whether it was to determine whether to move away or toward some location, or to change some process occurring within the cell.

"Chemical signals are released by signaling cells in the form of small, usually volatile or soluble molecules called ligands. A ligand is a molecule that binds another specific molecule, in some cases, delivering a signal in the process. Ligands can thus be thought of as signaling molecules. Ligands interact with proteins in target cells, which are cells that are affected by chemical signals; these proteins are also called receptors. Ligands and receptors exist in several varieties; however, a specific ligand will have a specific receptor that typically binds only that ligand."[88]

As life evolved on Earth, many other forms of communication began to develop. Specialized organelles began to form, like flagella which are used for locomotion in early microorganisms that began to be utilized for communication. Early multicellular ACTs likely utilized both appendages like antennae, as well as organelles capable of producing electric currents, in order to communicate and sonar. While marine life may not communicate directly with each other using these signals, they are a type of sensory organ that communicates information about their surroundings. Being able to sense the minute changes in nearby electromagnetic fields is vastly superior to chemical gradients because they travel at or near the speed of light, depending on the medium. Touch-based forms of communication are also possible, as is the case in organisms using antennae and appendages to communicate with one another through various motions in these appendages.

We even see this in land insects of today, where ants greet each other and use their antenna to communicate. Interestingly, bees utilize both scent-based communication as well as pattern-based communication in the form of dances.

"There are 2 types of bee dances: the round dance and the tail-wagging or waggle dance, with a transitional form known as the sickle dance."[89]

These dances describe general location and proximity, as well as food quality to other bees. This is vital information for the survival of the colony since it would be the case that if these bees could not communicate with one another about their findings, the entire colony would likely starve.

In more modern sea organisms, like sea mammals, we see sound-based communication utilizing vocal structures that are all similar to the same system that humans currently use. While these are also specialized organelles, the differentiating pattern here is that the communication exits the individual in the form of a wave and enters the other organism through another organelle. It is not as simple as a single organelle interacting with the same organelle of another organism.

With further evolution, BCTs began to inscribe abstract figures on the walls of caves. This type of communication represents the first time that the *fungus* had iterated toward a form of indelible (or semi-permanent) communication that allowed for information storage over long periods of time. These pictures don't communicate specific words or phrases but a generalized impression of a condition or situation that these early BCTs experienced. Fast forward to around 3500 BC and the ancient people of Sumer were using the first known written language. This is an extension of the abstractions of earlier BCTs that allowed for more precision of communication as well as a faster way of communicating specific information.

IPT As the BCT evolved into modern-day humans, language flourished. With the improvements in language, the rate at which BCTs were able to learn and innovate increased significantly. This increase in the rate of learning allowed for an increased rate of technological advancement. Modern-day humans are capable of using all modalities of communication in the fungus that existed prior. If someone is deaf or mute, they can use sign language. If someone is blind, they can use braille. Otherwise, we can use vocalization-based communication.

In addition, with the advent of the early computers discussed above, humans have forms of communication that are electromagnetic in nature. During World War II, the development of global transmission resulted in communication networks that allowed individuals to be made aware of what was going on at any place around the globe within minutes to hours instead of weeks to months. Prior to the advent of these technologies, films of the battlefield were taken and sent back to show Americans at home what was actually happening out on the field.

Today, the edge in communication on the battlefield comes from satellites, the internet, and the computers and laptops connected to them. Utilizing these technologies, we can transfer information in forms that we understand into digital signals, which then are communicated between members of the Technium (networked computers or sender/receiver transmission towers), and then converted back into signals and symbols that humans can understand. In essence, humans have leveraged multiple different offshoots that occurred in the evolution of communication in the fungus in order to come up with a revolutionary and novel way to communicate. This method of communication can cover extreme distances and occurs at literally lightning-fast speeds.

Interestingly, and to further the point on the importance of efficiency and speed of communication, English is not an original language. It is a derived language that is a hybrid of many other languages. As English has evolved over many hundreds of years, it has become further refined and adapted toward promoting more information-dense transfers while communicating. This is why English is one of the most used languages today. It is very important to get the most meaning across as fast as one can in order to save lives, as well as to accelerate learning.

"University of Lyon researchers François Pellegrino, Christophe Coupé, and Egidio Marsico didn't travel the world, nor did they survey every single language, but back in 2010, they did...determine the speech information rate of seven of the world's most spoken languages: English, French, Italian, Japanese, Spanish, Mandarin, and German. English came out on top, but not by much. Most of languages grouped pretty closely together, however, Japanese lagged behind the rest."[90]

Growth through communication can also be seen in the lyrics of popular music. Music, especially that of the more modern era, is one of the best ways to express sexuality and individual growth that often bypasses the restrictions of societal norms, cultural barriers, and even governmentally imposed restrictions. It once was said, "The quickest way to undermine any society is through its music," which was in the context of degrading or destroying that society's music. Conversely, bolstering expression and fostering new ideas through music creates a more robust society with greater levels of understanding and empathy, which promotes its survival by easing tensions and reducing conflict.

The result of more people being able to transmit more information at higher speed is the primary mechanism that drives conscious-growth.

This demonstrates the fungal process by which information transfer has evolved throughout the course of the development of life on Earth, as well as its importance for the survival of organisms as they employ these ever-advancing communication techniques. Communication will need to continue to increase in its information density and transmission rate as the Earth heats up so that complex problems can be shared, understood, and solved with the advanced tools of the Technium.

BP These developments in communication as the *fungus* has progressed are essential to the development of the fungus. It is very likely that each arm of the fungus is communicating with the greater whole as many of the interactions, which could be considered communications, between various plant and insect life forms are crucial support systems for the entire biosphere.

"Bees are not only extremely important for humans, but also for entire ecosystems to function. As we know, bees allow plants to reproduce through pollination. These plants contribute to the food system by feeding animals – aside from humans – such as birds and insects. If the food source for these animals was diminished or lost completely, it would cause the entire food chain to suffer. Additionally, about 80 percent of flower plants depend on pollination. If this process stops, not only do we have the potential to lose beautiful plants, but also food for us, birds, squirrels, and all of the other animals that depend on plants for food."[91]

One could only imagine the importance that many other seemingly insignificant organisms in the connected fungal network of life on Earth hold. In fact, this is the way the fungus progressed throughout all of Earth's history and will continue to progress as life continues to evolve. The awareness of these extremely interconnected sets of networked communication and feedback systems that exist between all life forms of the fungus demonstrates that the BCT is totally reliant on everything that led up to it. Humans will need more time to evolve but are on the path to truly understanding how important each and every organism has truly been.

After this further evolution, the advancements made in sending our thoughts and communications digitally may be made irrelevant as we may begin to tap into telepathic abilities, owing much to this understanding of interconnectedness. This could also increase the speed of communication many times the current rate at which humans process thoughts and information. For example, people begin to struggle to understand the information presented to them when it is interpreted through spoken word that is sped up more than one and a half to two times. Some people have no trouble with this, as you can speed up videos on YouTube to this speed. However, others listening to the same information will be unable to understand it at all.

From SCCTs to BCTs on the verge of achieving TES, the fungus primarily iterates toward improvements in communication as described previously. Each of these advancements is a natural and subsequent result of the small iterations and could likely not have been achieved otherwise. By the time humanity is ready to leave the Earth, it will have achieved levels of communication efficiency and transfer rate that are unimaginable at the current time. As mentioned before, they will be necessary in order to ensure the survival of TCMs as the amount of information, as well as the problems that require solving, will balloon the current transmission rate and data volume requirements. For example, a single strand of DNA digitally stored on computers takes up approximately 750 MB of data storage. However, opting to use DNA over digital storage systems can allow for 215 PB (1 petabyte = $1x10^{15}$ bytes) in a single gram of genetic material. Currently, the largest usable drives that are available to the average consumer are in TBs (1 terabyte = $1x10^{12}$ bytes) and weigh hundreds to thousands of grams. This demonstrates that 1 gram of genetic material is on the order of one million times more efficient. With the advent of quantum computers, which operate on atomic-level or quantum principles and can compute far faster than traditional computers for some problem types, it is apparent that leveraging atom-sized storage technology, such as the DNA storage mentioned above, will unlock both the necessary computational speed and informational density required in order to achieve TES as a TCM.

BBP Throughout the universe, fungal growths will sprout up wherever the conditions of life support them. In each of these growths, communication is an essential driver of the evolution of that fungus toward TES. Each growth will require faster and more information-dense forms of communication in order to ensure the survival of that growth. Some growths may adopt different communication styles in the BCT phase, or even in the Technium that they develop. However, the general motif will be the same. Without these advances in communication, the fungal growths will never achieve sufficient complexity or speed of informational flow in order to surmount the task of leaving their biospheres. Thus, one can consider that one of the elements of achieving MCG that is promoted by the universe is of ever-advancing communication and that this improvement in communication is a key element of fungal growth throughout all life in the universe. By each fungal growth learning of other fungal growths in the cosmos and sharing knowledge via learned communication modalities, the universe brings consciousness one step closer to the UGCH.

It's Always Been about Transportation

TL;DR: The essay explores the concept that transportation is the ultimate survival need for the fungus, highlighting the evolutionary journey from microscopic tails in early life forms to advanced modes of transport like aerospace. The focus is on how mobility has been a key factor in the survival and evolution of life, driven by the need to seek nutrients and avoid threats. The narrative connects the evolution of transportation, from early organisms to modern advancements, as a fundamental aspect of life's quest to avoid inevitable death and move towards post-terrestrial concerns.

From microscopic tails wagging in early single-celled and multicellular life forms to the fastest swimming fish, becoming amphibious then quadrupedal, to becoming bipedal, from humans riding horses to horseless carriages (automobiles), from the car to the airplane, and, finally, the airplane to aerospace – it's funny to think that this is what it's always been about: transportation. All life (the *fungus*) is attempting to avoid the second limitation of the universe: inevitable death. When consciousness increases and awareness expands, the focus of the fungus shifts from terrestrial concerns to post-terrestrial concerns. This is only possible due to the ability of the fungus to be increasingly mobile.

Early Single/Multicellular Life (SCCT/MCCT)

Early organisms on Earth (archaea, eukaryotes, and bacteria) would exist in a much different world than the one we live in today. Bereft of the many predators that exist that current animals and insects must avoid, responding to their environment was a key driver in the evolution of these organisms. If nutrient density in the vicinity of one of these organisms should decrease, evolution would favor an organism that could move in order to seek out sources of nutrients rather than those that wait around for conditions to improve since they lacked the ability to move. While the true origins of the ability of these early life forms to begin moving are unknown, what is known is that flagella (tail-like structures) formed in order for these organisms to gain the ability to move through their environment. These flagella are developed using a class of proteins called motor proteins. Kinesin is one such motor protein utilized by eukaryotes for shuttling vesicles of nutrients across cytoskeletal filaments.

"Motor proteins use energy in the form of ATP to "walk" along specific cytoskeletal tracks. They are essential for movement of vesicles and other cargoes within cells, as well as for the movement of muscle and cilia/flagella."[92]

This was the dominant form of life for many millions to billions of years until sufficient evolution of single-celled life had taken place. After single-celled organisms evolved to multicellular life, these changes to the structure of organisms were more related to respiration and cellular processes, as well as specialized sensing organs. No true significant advances were made in locomotor capabilities in this evolutionary step. It wasn't until organisms with spinal structures (vertebrates) began to evolve that more significant changes in locomotor capability began to appear in organisms throughout the biosphere.

Oceanic Organisms (ACT)

Just prior to the Cambrian explosion, a class of fungi called lichen fungi helped transform the Earth's atmosphere and paved the way for the rapid diversification of life that took place during the Cambrian Explosion.

"Lichens are believed to have been the first fungi to team up with photosynthesizing organisms like cyanobacteria and green algae. Lichens can live without rain for months, providing protection for photosynthesizing organisms, which produce oxygen and release it into the atmosphere."[93]

The life forms that either diversified or evolved anew during this time period were fungi, arthropods, and fish. As these organisms began to evolve and spread throughout the oceans, some of them were doing so by utilizing an evolved ability to move around with appendages that were more powerful than the flagella of earlier multicellular and single-celled life forms. It is likely that as multicellular life became larger and denser, it would sink to the bottom of the ocean. Thus, flagella were no longer a reliable appendage for locomotion. These larger creatures formed structures that would allow them to move through the oceans with greater ease. While arthropods grew appendages that resemble legs, they did not develop a spine, which will turn out to be a pinnacle development for life on Earth. Fish, however, did end up developing a spinal structure and utilized this spinal structure in conjunction with a tail that is connected to the spine in order to generate extremely quick and efficient locomotive capabilities.

Over time, various forms of aquatic life began to form and thrive in the oceans. This increased competition for survival in the oceans ultimately became a threat to the life forms living in the ocean. What once was a haven for the early life forms to feed upon and thrive, became a hostile environment that threatened their existence. The predator/prey relationship between the many organisms living in the oceans forced some of these fish to leap out of the water to escape predators. Eventually, even flying fish, which were first seen in the fossil record, evolved hundreds of million years ago.

"Scientists analyzed fossils they excavated from southwest China in 2009. The ancient bones come from a marine fish named Potanichthys xingyiensis…The fish lived about 235 million to 242 million years ago in what researchers call the Yangtze Sea. This was part of the eastern Paleo-Tethys Ocean that sat about where the Indian Ocean and Southern Asia are now located."[94]

These fish were following the same imperative the fungus seems to have in every iteration: how to improve the ease and efficiency of motion. While this was developed over the need to escape predators, even if only for a few seconds to minutes, it demonstrates that the model that early single-celled life forms had developed is still relevant to future life. The need to move more efficiently and over greater distances is still programmed into the evolution of all life forms.

This particular evolution in fish also demonstrates that a certain level of competition and threat of predation in the oceanic ecosystems (inevitable death) forced fish to escape these threats by adapting to an environment outside the confines of the water: they became amphibious. Amphibious creatures were then able to gain access to more resources without the same level of threat as animal life on land had not yet evolved. Thus, this competition was essentially non-existent on land.

At each level of fungus development - ACT,QCT, BCT - we can see an increase in opportunity of civility. Here is a good vision of the growth of the *fungus* as it is actually trying to leave Earth.

Land Animals (QCT)

With the continued threat of predation and starvation in aquatic life, amphibious organisms began showing up in larger numbers as well as began acclimating to dryer conditions over millions of years. During this same time, approximately 450 million years ago, the spreading of land plants prepared the surface of the Earth for the greater numbers of animals that were beginning to appear on land. These plants served as a food source, as well as a source of protection from the Sun, as amphibious life began to take root on the surface. Over many more millions of years, the first land animals that were no longer capable of surviving underwater began to appear.

The plants that began to spread throughout. Earth's surface during this time aided in shifting the atmosphere. The conversion of carbon dioxide to oxygen is part of the photosynthetic process, which influenced the early amphibians to begin preferring the energy content of gaseous oxygen over the dissolved oxygen in aquatic environments. Due to this shift, organs that were better adapted to extracting oxygen from a gaseous environment began to evolve, which ultimately ended up in the formation of lungs. Lungs, allowing for greater uptake of oxygen, also allowed for greater efficiency in energy production and storage. This increase in efficiency allowed for the entire structure of creatures to be further explored by evolutionary mechanisms. Larger relative brain size, elongated limbs with more intricate muscle groups, and changes to sensing organs-all occurred at this point. Each variation that occurred was a necessary step the fungus must have gone through in order to iterate to an optimal design.

One such design that was optimal for quite some time was that of quadrupeds. Land animals that moved on four legs began to crop up. Amphibians were some of the first quadrupeds (appearing approximately ~390-360 MYA), but later reptiles (appearing ~315

MYA) and prehistoric primates (appearing ~178 MYA) began to evolve. These animals were very efficient at traveling long distances either through the trees or on land and were well-adapted to their environment. Through many more millions of years, and the impact of a meteor that wiped out the dinosaurs (~66 MYA), a massive shift in the ecosystem took place.

Organism	**First Fossil Appearance**
Amphibians	360-390 MYA
Reptiles	315 MYA
Mammals	210 MYA
Prehistoric Primates	178 MYA
Dinosaur Extinction	66 MYA

This allowed room for primates (such as Purgatorius mckeeveri) to flourish as they were not under the constant threat of dinosaurs. As these primates left the trees and began exploring the land, their bipedal capabilities began to increase. Evolution was also experimenting with changes to the appendages used to grab tree limbs as trees became less abundant and open grassy fields began to become more prevalent. As these early apes began to experience a greater distance between trees, their ability to run from predators required that they stand for long periods of time and be able to run faster in order to survive, eventually standing permanently. At this point, the next massively successful form of life had evolved, which would eventually outperform the quadrupeds.

Upright Creatures (BCT)

The apes that were able to hold their offspring while running and defend themselves with tools made by precise manipulation of their surroundings would go on to become more successful than other apes. These evolved apes (hominids) were the first bipedal organisms to inhabit the Earth. The success of these creatures comes from the fact that with two of the appendages of quadrupeds now being liberated, many evolutionary changes could take place in order to create arms and hands that were much more efficient at manipulating their surroundings. Over time, these hominids began crafting technologies to transport them over the surface of the water so that intercontinental travel became possible.

Hominid Species	First Fossil Appearance
Ardipithecus ramidus	4.4 MYA
Australopithecus afarensis	2.95-3.85 MYA
Homo erectus	1.5-1.9 MYA

A. Early Boats/Rafts (9–900 KYA)

There is possible evidence that the earliest known trips made at sea occurred nearly a million years ago. The evidence comes from the fact that multiple migrations were supposed to take place in different groups of early humans at various times throughout history. It would be difficult to explain the appearance of various cultural, technological, and agricultural advances throughout the world if some form of migration that would likely only be possible by sea travel took place during the development of early human societies.

"...it's apparent that H. erectus--which may have survived in Java until 30,000 years ago--launched the first age of ocean journeys between 900,000 and 800,000 years ago. On Flores, an island separated from Bali by ocean waters and the islands of Lombok and Sumbawa, other scientists have dated stone tools at more than 800,000 years old…A quantum leap in cognition and technology occurred around 900,000 years ago…All the traits that fundamentally define modern humans were first developed by Homo erectus."[95]

While these suppositions have sound logical support, they are not sufficient to conclude a known fact about the earliest sea travels. True concrete evidence for aquatic vehicles comes from excavations of artifacts that date back to around 7,000-10,000 BCE. This canoe is now stored at the Drents Museum of Art and History in Assen, the Netherlands.

"Named The Pesse Canoe after its location, this is the oldest known boat in the world. Found by a local farmer in 1955, this prehistoric vessel was made about 10,000 years ago…It dates back to the Mesolithic, and it is estimated it was built somewhere from 8040 and 7510 B.C.E."[96]

These devices for transportation were made in order to carry supplies or escape hostile environments. Once again, the threat of predation and starvation drives the evolution of transportation in the *fungus.*

B. Domesticated Travel (~7 KYA)

While seafaring travel was very efficient for Homo sapiens, the need for more efficient land travel still existed. Around 7 kya Homo sapiens began domesticating oxen and horses for agriculture and transportation. These animals showed up just prior to early humans in Earth's history. This use of animals allowed for greater ease of both broadening the reach that Homo sapiens had as well as transforming their environment to suit their needs. This is the first time in history that an organism had a deliberate effect on the environment around it in a planned and systematic fashion at a large scale. The hardships of plowing the land and harvesting crops influenced this behavior as these were much more difficult tasks for Homo sapiens to carry out by themselves. The natural increase in conscious awareness that the extra limbs afforded Homo sapiens allowed them to come up with unique solutions to problems that may be solved by the use of other creatures. This trend continued for thousands of years and still continues to this day. It was this advancement in transportation that opened the door to future advances in transportation both assisted and unassisted by animals.

C. Wheeled Vehicles (6-7.3 KYA)

The first use of the wheel was for aiding in the creation of pottery, which occurred just before the advent of its use in transportation.

"Thanks to these finds and others, scholars agree that the wheel was invented somewhere in Eurasia or the Middle East between 5000 and 3400 B.C. Once developed, the technology spread quickly, transforming society as it went."[97]

The oldest wheeled transportation device we have archaeologically discovered, which is not to say earlier discoveries cannot be made, is the chariot. Chariots were used for multiple purposes, however, they later became primarily used for hunting and warfare as their use became more widespread. As with most technologies, the initial use case is usually not the final use case, as it takes some time for society to adapt to all the possible uses of new innovations. The wheel is one such example of a highly versatile technology that has a plethora of use cases.

"...open, two- or four-wheeled vehicle[s] of antiquity, probably first used in royal funeral processions and later employed in warfare, racing, and hunting. The chariot apparently originated in Mesopotamia in about 3000 BC...On the earliest chariots the wheels rotated on a fixed axle that was linked by a draft pole to the yoke of a pair of oxen. To the axle was attached a superstructure consisting of a platform protected by side screens and a high dashboard. These Mesopotamian chariots were mounted by both spearman and charioteer...."[98]

The wheel, and its use in transportation, was a tipping point for the development of civilizations around the world. Improvements to the wheel, which included making it lighter and using spokes instead of a solid structure, opened up its use in many applications. This could be considered one of the pivotal moments in history that allowed for the true burgeoning of what modern civilization would ultimately become, as well as spawned greater technological advancement as the necessity to transport more people and materials grew.

D. Steam Engines, Electricity, and the Internal Combustion Engine – The Industrial Revolution

Steam engines were developed to increase power output in comparison to travel based on the use of animals or other manmade structures. They were also used as pumps in their first iterations. Around this time, other technologies were being fine-tuned and improved and would ultimately come together to help fuel the Industrial Revolution. In addition to steam-powered engines, the other technologies being implemented during the Industrial Revolution were combustible fuels based on petroleum and the transfer of electricity. Most of these technologies relied on either coal or petroleum products as their starting point and could convert the chemical energy stored in these substances into other forms, as is the case with the steam engine. The steam engine would utilize a burning lump of coal to heat water and convert it to steam in order to generate locomotion.

The first steam engine used for transportation was in 1785-1787, where American inventor John Fitch began adapting the steam engine toward use in boats. Fitch demonstrated his boat to congress in 1787 but the steamboat didn't gain wide adoption until later.

"Steam engines first appeared in the mid-1700s, often modeled after James Watt's successful 1769 patent. American inventor John Fitch looked to use this new

invention as a propulsion device for a new method of transportation. Beginning in 1785, Fitch set about building a steam powered boat, completing his 45 foot long prototype two years later and successfully demonstrating it before members of Congress in 1787."[99]

As the steamboat became more widely known, other applications of steam began to crop up around the same time. The world's first steam-powered passenger railroad was the Stockton & Darlington Railway, which was originally constructed as a proof of concept in order to take advantage of a large vein of coal in Darlington.

"On September 27, 1825, the first engine ran from Darlington to Stockton, preceded by a man on horseback carrying a flag reading Periculum privatum utilitas publica ("The private danger is the public good"). When the horseman was out of the way, Stephenson opened the throttle and pulled his train of wagons carrying 450 persons at a speed of 15 miles (24 km) per hour."[100]

Shortly after the boom in locomotive technology that steam provided, other inventors were applying this technology to advance personal transportation. Francis and Freelan Stanley were two inventors that developed the steam-powered car in 1897. Not long after proving that their invention was feasible, they formed the Stanley Motor Carriage Company.

"They built their first steam-powered car in 1897 and sold more than 200 by 1899, making them the most successful U.S. automaker at the time."[101]

Around the same time, Karl Benz commercialized the production of internal combustion vehicles, which were invented a few decades prior, in 1886. These two inventions, the steam-powered vehicle and the internal combustion engine, transformed the world in that they allowed the mass adoption of technology for personal transport. With the advent of these technologies, individuals became free to travel long distances and at greater speeds, as well as transport large amounts of material over these large distances. The most successful, and well-known, of these early internal combustion engine vehicles would come from Ford in the form of the Model T. The Stanley Steamers were quite expensive, whereas the Model T was around 7 times cheaper. Ultimately, it was this low-cost option that allowed transportation to become truly personal and available to the masses.

E. Air Transportation (Hang Gliders - 1890s / Wright Brothers 1905)

Air travel would be the next natural extension of transportation technology that humanity would turn to after success with transportation on land and in the waters. By the mid-1800s, a British engineer by the name of George Cayley had researched aeronautics and created the first glider. His initial success paved the way for further improvements. While his invention's first flight was a milestone in and of itself, it ended in a crash landing.

"Cayley established the modern configuration of an airplane as a fixed-wing flying machine with separate systems for lift, propulsion, and control as early as 1799... In 1804 he flew the first successful glider model of which there is any record. His work culminated in 1853 with the completion of a full-scale glider that carried his reluctant coachman on the first manned glider flight on record."[102]

Later, Otto Lilienthal tinkered and improved upon the work of Cayley and produced a monoplane hang glider in 1894 that was to be one of the most influential pieces of tech prior to the first airplane flight.

"Lilienthal made some 2,000 flights in at least 16 different glider types between 1891 and 1896. The largest number of flights were made with his standard glider, a monoplane aircraft with a stabilizing tail at the rear and wings that resembled 'the outspread pinions of a soaring bird'."[103]

At or around the same time, the Wright brothers were tinkering with Lilienthal's kite design and improving upon it.

"Their first experiments with 'wing warping,' as the system would be called, were made with a small biplane kite flown in Dayton in the summer of 1899. Discovering that they could cause the kite to climb, dive, and bank to the right or left at will, the brothers began to design their first full-scale glider using Lilienthal's data to calculate the amount of wing surface area required to lift the estimated weight of the machine and pilot in a wind of given velocity."[104]

After iterating on the kite design, which ultimately became the biplane design that we are familiar with, and trying to improve upon issues such as lift and general control of the aircraft, Orville and Wilbur Wright created a controllable and sustained flying aircraft in 1903. A couple of years later, in 1905, they created the first fully practical airplane.

In the 100-120 years that followed these initial breakthroughs in aviation, many rapid advances were made. The first power-controlled airplanes could hardly go a few miles; but by WWI, they were used for short-range bombing missions. In the time from WWI to WWII, the plane had a much more pivotal role in warfare. Most of WWII was fought with planes, with the Royal Air Force making it all the way from Britain to Germany and vice versa during the blitzkrieg bombings.
In the post-war era, due to the advancements brought about by war itself, airplanes could travel for thousands of miles. To demonstrate the importance of traveling further, the reason PanAm (Pan American Airlines) named itself as such is because, at the time of its founding, being able to go from coast to coast was an impressive feat. Similarly, one of the reasons Howard Hughes is considered a genius is that he was one of the first to figure out how to have a cross-continental flight from the United States to Europe, something that has only gotten faster and cheaper since then. As mankind had now tamed the land, sea, and air, the next logical step was to begin to address the reaches of outer space, starting with our own extraterrestrial backyard.

F. Space Transportation (1961 - Russian Cosmonaut)

The requisite technology for rocketry had existed since at least the 13th century when the Chinese were experimenting with gunpowder-filled tubes. These gunpowder-filled tubes were adapted to be used in battle when the Chinese and Mongolians were at war. Later, the foundation of rocketry was laid by Sir Isaac Newton when he organized his understanding of the laws of motion into formulations and equations that could accurately predict and describe the way objects moved under the influence of gravity. Later, steam-powered rockets were attached to automobiles and small rockets were tested on the ground.

Colonel William Congreve was intrigued by the use of rockets by the Indians against the British in 1792 and 1799. Congreve then committed to making rockets for the British military. These rockets were very successful and were utilized in the war of 1812. They are also a notable mention as they appear in the lyrics of the Star-Spangled Banner when we sing, "the rockets' red glare", something every child knows but has no historical context to relate.

Congreve's rockets had not improved in accuracy much, as their success was more attributed to their numbers. Not much later, rockets were outperformed by artillery and were relegated to being used in festive activities rather than war.

It wasn't until 1898 that a Russian schoolteacher by the name of Konstantin Tsiolkovsky proposed the idea of rocketry for space travel. He released a report in 1903 that advocated for the use of liquid fuels in order to increase the range of these rockets as he related the range of a rocket to the speed of the escaping gasses. Later, in 1919, Robert Goddard released a pamphlet that described the process of reaching extreme altitudes with rockets. As Goddard began experimenting with rockets, he was using solid rocket propellant, but later became convinced that liquid propellant was necessary, which was in agreement with Tsiolkovsky's earlier findings. With further experimentation, Goddard invented both gyroscopic controls as well as parachutes in order to stabilize flight and safely return the rocket back to the ground. Goddard is considered the father of modern rocketry.

Hermann Oberth, another pioneer in the advancement of space flight technology, released a book about space travel in 1923 which went on to influence a group of scientists interested in advancing rocket technology with the hopes of reaching outer space to create the Verein fur Raumschiffahrt (Society for Space Travel). Most notably, Wernher von Braun headed the production of the V-2 space rocket as a member of this society. This rocket was used against London in World War II.

Once again, war promoted and accelerated the advancement of technology through the second fundamental law of the universe (inevitable death). Both with Congreve's rockets and in World War II, the rocket was improved in order to become useful as a more advanced weapon. It wasn't until a new kind of war, a decade after the end of World War II, that the rocket would be advanced again in order to bring mankind into space. This was the "Space Race" against the Russians. While not a standard war in the sense of battles with large death counts, this technological race determined who would be the greatest nation in the world. Whoever the victor was would dictate the course and trajectory of the evolution of our species for decades to come.

The Soviets scared America when they announced their first successful orbit of the Earth with the launch of Sputnik in 1957. Not long after, the United States launched its satellite, Explorer I, into orbit in 1958.

Brief History of Rockets

Not long after this first achievement, the Apollo Space Program was created with the intention of launching a man into space. Wernher von Braun led this program after having surrendered at the end of World War II. In 1961, the Russians scared the Americans once again with the news that the first man in space, Yuri Gagarin, aboard the Vostok 1 had entered orbit. In 1968 the first Apollo mission was launched and shortly after, the first moon manned mission was launched in 1969. It was John F. Kennedy's declaration, as well as a call to other nations, that we would make it to the moon by the end of the decade in the 1960s, that pushed NASA (the National Aeronautics and Space Administration) to create the machinery and train the astronauts that would put America first in the "Space Race." This was achieved by Apollo 11's landing on the moon in July 1969. Further missions for visiting the moon were carried out, but this was the major milestone that determined the victor of the "Space Race."

What Was the Apollo Program? | NASA

As a natural consequence of exploring space, the idea of living in space comes next. The first crude space lab came about in 1969 when NASA launched Skylab. This was the first livable environment outside of the confines of Earth's atmosphere. However, the third fundamental limitation of the universe is that life is biospherically-bound. Thus, these space stations would have to recreate the environment that exists on Earth for those that live in these stations in space.

The International Space Station (ISS) has been in orbit since 1988. Humans have been living in space for over three decades. Projects like Space X are taking existing space travel technology and making them cheaper and lighter while carrying a larger payload and being reusable.

Skylab: America's First Space Station | NASA
History and Timeline of the ISS

BP At each stage of the *fungus*' development, there seems to be an antecedent development or developments that fit perfectly together in order to catalyze a truly transformative evolution that becomes an essential and vital piece of the fungus' survival. Are these merely providential occurrences, or perhaps something more structured or intended? This set of occurrences is the fungus generating what it needs in order to grow and transport an organism of greater consciousness and continue iterating toward a spore. Starting with the earliest life forms and progressing towards intercontinental ships of ancient times, the fungus progressed significantly in its ability to generate movement in various conscious-growths. From that point, advancement has been even more rapid. It only took Homo sapiens sapiens (modern humans) sixty-six years to go from the first controlled flight (the Wright brothers in Kitty Hawk, NC) to Neil Armstrong's first flight to the moon, whereas the first tools that were developed by early hominids like Homo habilis were used for millions of years before any significant change occurred in their usage and design. In the modern era, multiple innovations related to energy transport and storage were all independently occurring and making improvements in their respective domains. However, they would ultimately come back together in order to generate transformative technologies such as the steam engine and the internal combustion engine. This progression demonstrates the fungal nature of technology and innovation by first separating, for specialization, and then merging, to generate truly transformative technology. Initially, when a fungus spreads its networks from the central point of growth, traversal occurs slowly. However, when the fungus has spread considerably, the new ground that can be covered and scavenged increases exponentially. We now have spaceships that have taken modern humans to the moon and will eventually do the same to Mars. While these starships are in their infancy, they are profound achievements on the path to TES.

While conflict and war have been some of the strongest promoters of technological growth, endeavors into space are becoming increasingly important. Many of our daily technologies initially started off as space-age technology developed by NASA that was first used in space programs or were a direct consequence of research carried out in space. Thus, as the fungus continues to push forward toward achieving TES, the importance of both surviving in and traveling through space will only increase and contribute ever more significantly to the body of technological capabilities that humans possess. At some point in the near future, the main focus of the fungus' spore (humanity) will be solely focused on two things: generating technologies that can assist in the survival of the species on Earth, as it develops the technologies required to survive space or technologies that will allow them to survive in space in a sustainable manner. If eyewitness accounts and reports of

Unidentified Aerial Phenomena (UAPs) are to be trusted, which have been released by members of the armed forces in greater numbers in the last decade, we have evidence that other conscious-growths have achieved the very thing being discussed here.

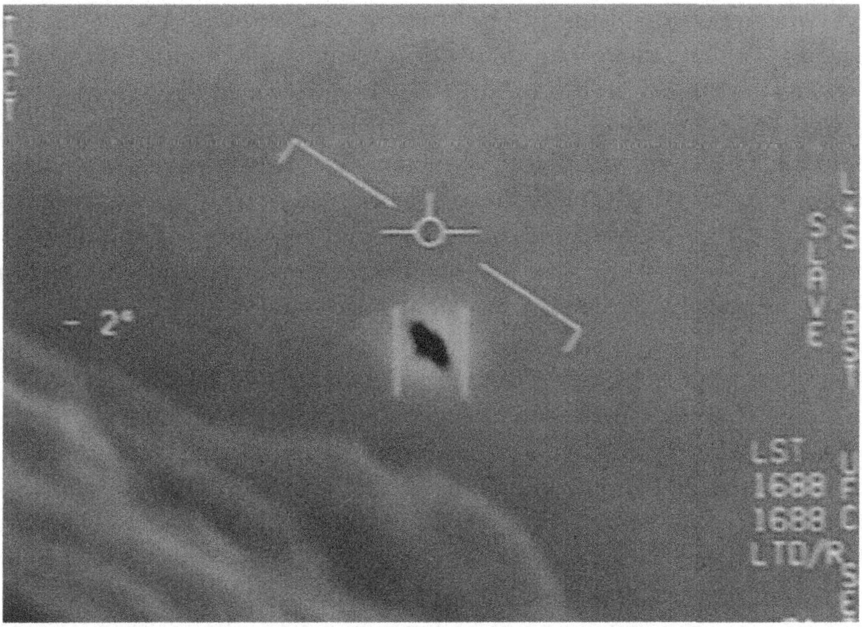

"Image of a UAP that was reported by a US Navy Fighter Jet Pilot serving on the USS Nimitz. UAP appeared on both radar and visual systems. Alex Dietrich and co-pilot David Fravor were sent to investigate anomalous radar readings off the coast of Southern California. The pilots admit that this object accelerated in ways they have never seen and could not explain."

Should these sightings be of extraterrestrials, it is clear that they have achieved TES and are exploring the cosmos, potentially watching the development of technology on these biospheres, and in some extreme cases, influencing them. These devices could be probes, acting simply as a tin can moving through the cosmos with either a pre-programmed function or being controlled remotely. It is also possible they house beings (TCMs) within the tin can. If they are able to make it to our planet and execute the maneuvers that have been proposed in their sightings, it is quite obvious that continual development of transportive technologies is an absolute must.

In the end, Earth is a temporary spawning ground for the *fungus* and the conscious-growths that begin to evolve on it become increasingly aware of this fact. During this initial phase of organic evolution, a variety of different locomotive methods and technologies form in the fungus. As this occurs, a shift in focus takes place in the more consciously aware and concerned organisms at the forefront of the fungus: the BCTs. It is these BCTs that will end up innovating and inventing until TES is achieved and transportation technology has advanced enough to develop the tin can for the "tin can man" to live in ESS. Ultimately, every advancement, whether it be medicine, communication, material processing, or agriculture, is for the improvement of transportation. In some way, these technologies will all merge back and push the fungus forward in order to enhance the locomotive capability of the fungus to create the possibility of post-terrestrial existence.

BBP Throughout the cosmos, the process of matter expanding and forming clusters has been occurring since the very beginning of its expansion. However, the nature of star birth and death has demonstrated that certain elements that promote the development of life were born in the very first stars; and thus, there could have been no life developing prior to the development of these elements by the first stars. Similarly, the expansion of the universe also dictates some limitations on the development and maximization of conscious-growth throughout the universe.

Since the universe continues to expand faster with greater distance from any given point within the universe, there will come a time when galaxies and star systems are so far removed from one another that the light from one region can no longer make it to another region. Couple this with the fact that the number of stars being born throughout the universe is decreasing rapidly, and it is estimated that there are only a relatively small number (5%) of stars that will ever be born from this moment forward, and it becomes apparent that there is a timeline for which life can appear and consciousness can evolve and grow.

These facts demonstrate that there is a system in place for growing consciousness under the same threat as discussed prior: inevitable death. Conscious-growths that appear before the end of this timeline will continue to be influenced in ways that follow the dynamics discussed prior. Their ability to see beyond their own star systems and into their galaxies and beyond is what influences future awe and wonder. These psychological effects promote continued technological advancement and make these conscious-growths aware of the need to develop technologies of transportation to aid them in surviving the death of their birthplace. However, the life that spawns at the end or after this timeline will not be exposed to the same influencing factors. They may see their star and a few surrounding planets. However, they will not be able to see a galaxy or other galaxies floating through the cosmos in these far reaches of time. It is these later fungal developments that are much more likely to fail due to their lack of motivating factors that affect their psychology and inspire any sense of awe or wonder. The only threats they will be faced with are those of their local star systems.

Thus, fungal growths are all about transportation; however, the prime period of time in which fungal growths can be influenced to generate these technologies is a function of the physics of the universe itself. Fungal growths that are of sufficient conscious awareness and exist before the period in which these influences begin to disappear will be acutely aware that their time for being able to determine which direction to travel in order to explore uncharted regions as well as to study remote astrophysical phenomena is limited. This pushes these conscious-growths more frantically toward exploration and discovery, as the end of this "golden age" of development draws nearer to an end.

Conclusion

The next level of maturity (a MCS) of the *fungus* will have been achieved when humanity is no longer dependent upon the Earth for its survival due to technological advancement. At this point, a new epoch in human existence will have begun: the Galactiocene. Similarly, all conscious-growths throughout the universe will grow in the same way and be forced to escape their temporary circumstances, due to their impermanent home planet, in hopes of achieving permanence in the universe. This is a condition of absolutes for all conscious-growths that evolve throughout the universe. This reality may be made more relevant by knowledge of the fact that a star dies every 60 seconds, Inevitable Star Explosion (ISE) in the universe by going nova. In addition, and even more shocking, is that this the lower bound of the way in which the universe forces *fungus* conscious-growths out of their birthplace, as going nova is not the only process a star can undergo that would change the nature of habitability within a star system. Smaller stars may last longer but will ultimately reduce into dwarf stars ,which will pull the habitable zone of any star inwards, drastically shifting the nature of how life can be sustained within that star system.

It is the interplay between consciousness, awareness, and concern that drives humanity (spores of the *fungus*) forward via technological progression. Like single cells grow and evolve into multicellular life, a single invention along with multiple innovations generates technological growth that behaves like a fungus, showing that technology is part of the *fungus*.

This progression is fungal in nature as there exist many seemingly independent offshoots in technological advancement that end up being utilized by the fungus at a later date in order to achieve TES. As organisms on Earth became more conscious through the early evolution of their cellular machinery, the underpinnings of awareness were forming. As this awareness increased, concern began to creep into the mental framework of these more advanced organisms. Concern, as defined earlier, was an anticipatory behavior that allowed forecasting and prediction of future events and requirements. This ability stimulates the desire to avert the concern, which will ultimately drive technological advancement in sufficiently evolved organisms. The most prolific contributor to technological advancement on Earth by many orders of magnitude has been the BCT. The technological advancements that the BCT has achieved on Earth dwarf that of any other organism and this is due to its highly evolved brain. As technology progresses and conditions change, conflict is an unavoidable consequence of both limited resources and increasing civility and organization of organisms. Humans band together or fight wars over things that animals with lesser capabilities have zero idea or concern about. Thus, conflict enters into the calculus of technological advancement. Different groups of humans will be at odds with the beliefs they hold and experiences that they have endured due to differing circumstances in their local regions and cultures. Such differences drive conflict throughout the development of the spore. However, after the entirety of the fungus is made aware of and understands the temperature funnel and its implications, conflict is replaced by the adversity and stress that is continually imposed by the condition of absolutes within the universe.

Conflict's role in technological advancement is that it disrupts stagnation or the status quo. This has the effect of promoting technological advancements in two different aspects that are crucial and essential to the survival of the fungus, which are transportation and communication. Conflicts invariably require transfer of more information in order to assess situations, or more advanced encryption so that messages do not fall into the wrong hands. These requirements stimulate processing capability and informational density, which are essential for achieving the technological evacuation standards required to escape Earth's demise.

The other effect that conflict has on stimulating technological advancement is that of refining and revolutionizing transportive technologies. Whether it be the shipment of goods, the speed of combatants, or the ability to evade an oncoming threat, both terrestrial or extraterrestrial, the advancements made in transportation are the single most important contributor to TES and the primary focus of the *fungus*. Almost every kingdom of life has evolved unique and innovative locomotive capabilities as the *fungus* has evolved. Humans have continued to add to these modes of transportation by utilizing external technologies instead of biological adaptations. This is an absolute requirement to ensure that eternal self-sufficiency is achieved, for, without transportation, none of the other technologies required to maintain living conditions in the cosmos will have any meaning.

The bumps in the road along the way, such as the death of our star and the boiling of our oceans, only serve as further fuel for the bonfire of curiosity and the quest to fill the void in our collective psyche. For any fungal growth in the universe, this process would carry on in a nearly identical manner. The fungal growth would first grow in consciousness in order to finally become aware. This awareness would generate concerns that stimulate responses. These responses become increasingly technological, which drives the spore of the *fungus* toward achieving TES and promoting ESS while attempting to fill the void created in the psyche that formed within that spore caused by HTO.

5. PSYCHOLOGY

Introduction

This chapter will focus on the *fungus'* current state of conscious awareness on Earth, which has developed a bipedal organism (modern-day humans). The main purpose is to demonstrate the events that continue to take place in our daily lives, and for the foreseeable future, are a result of the developing psychology, which is influenced by a design mechanism that was baked into the universe at its inception.

TL;DR: The *fungus* spore's primary focus is securing its objective identity for ultimate survival. This spore is a product of the *fungus* to become the ultimate in surviving as an entity by securing its self-identifiable rights being reality or not. It will invent, construct, and kill to secure that identity, even if by killing itself. The analogy is drawn to the *Alien* movie's monster, emphasizing the spore's relentless drive to avoid death and re-identify with itself. This pursuit is linked to maintaining True Origin and contributes to the Maximum Conscious-Growth in the universe's construct. This section delves into the profound impact of the Void of Psychosis (VOP) generated by increasing awareness on conscious entities, particularly developing spores. It explores the concept of concrete identity, defined by the interplay between stored memory, sensory inputs, and the psychological condition of an entity. The analysis focuses on the ramifications of cognitive awareness of Hidden True Origin (HTO), leading to a void in concrete identity and subsequent psychosis. To address this void, entities engage in actions to signify their concrete identity, forming the basis for generating technology towards achieving Technological Evacuation Standards (TES).

Cinder Wall of Perceived Reality

This model represents the makeup of an individual entity or group's psychological structure as blocks of experiences that over time build a wall that is a perceived reality of that entity. Examples include: acceptance, abuse, recognition, discarded, profound experience. Depending on the Perspective and Perception Limitations (PPL), the psychology of each wall made during conscious evolution is a voicing of that entity's experience during that entity's time in a biosphere. Because of the VOP, each entity whether individual or group will perform actions trying to establish a perceived reality. To draw a sharper view of reality it takes all the Cinder Wall's combined blocks of experience developed through time to do so. Every generation of human growth has a 10% sharper view of what reality actually is. It is the subconscious (the *fungus* autopilot) that makes sure this phenomenon occurs. It is that generation's responsibility to be a contributor in this process making the future of humans a more survivable one. When an entity experiences new knowledge or understanding of new reality, they have the option to deny it or go into their Cinder Wall and rearrange their previously constructed blocks in a more organized manner. This sharpens that entity's or group's view of reality thus contributing to the collective's sharpening. No one wall is perfect or can be perfected as that would not contribute to the forward progress of conscious growth.

Let's look at a few examples of this process. An individual may learn something new about their lover that is not accepted well. That entity has the opportunity to deny the reality or go back onto their Cinder Wall and reconstruct a few of the blocks that may have contributed to the decisions that got them here. If a nation finds itself at war with another nation, it may be due to the fact that one nation is showing the other a newer view of reality, causing them to deny. However after one side loses the war, they have the chance to retro reconstruct their Cinder Wall into a better understanding of the phenomenon of conscious growth.

In applying the context of the material in the book, the Cinder Wall is a useful tool in applying new information and facts to a conscious entity discovered by collective consciousness as it grows.

CINDER WALL OF PERCEIVED REALITY

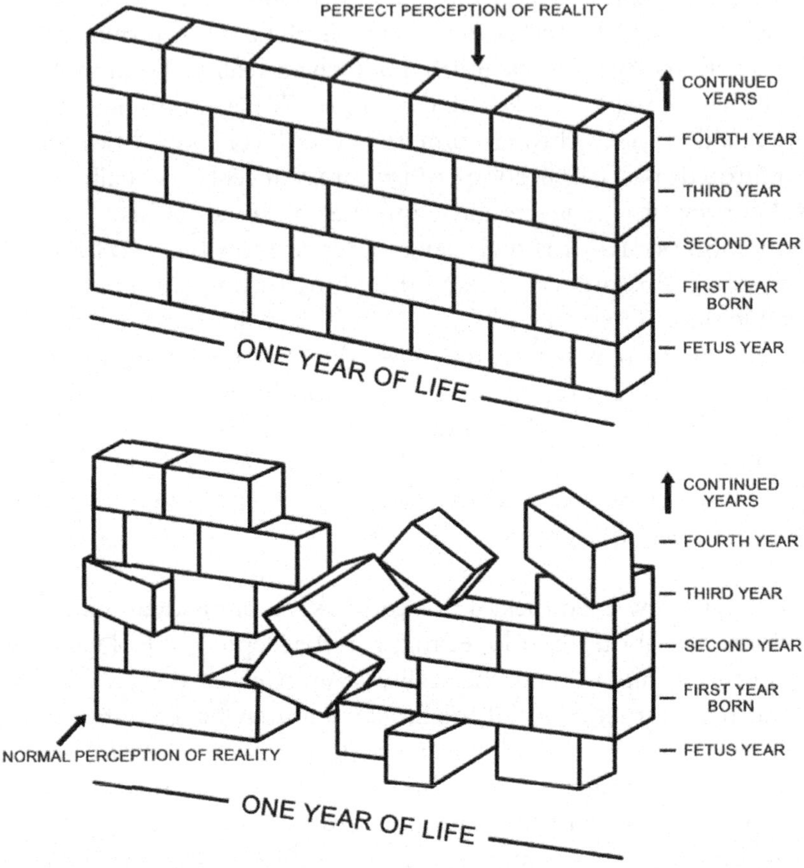

FIGURE C14

Figure C14: Cinder wall of perceived reality

This figure of a cinder block wall is a model for a physiological make up of an entity's experiences, input and knowledge. Each block represents a primary event or block of knowledge that forms a psychology measured in years for each level. This is a useful tool in rebuilding or correcting walls after receiving new knowledge. This may align the past and correct that psychology's current place in life more in tune with its reality.

Perspective and Perception Limitations

Perspective and perception limitations (PPL) are the key design in making sure that each individual conscious entity is a sole contributor to the collective consciousness or *fungus*. This setup defines each and every human independently such that the conditions and the self-interpretation of those conditions cause the entity to act and react. These actions are a subliminal voicing of the entity's experience backed by ancestral influences and fear and outcome of this entity's unknown future. The past accumulated growth of the collective is giving the perspective the entity needs to plow forward into its future. The success or contribution of this entity to the *fungus* is the message of the perception of the entity, whatever that perception may be.

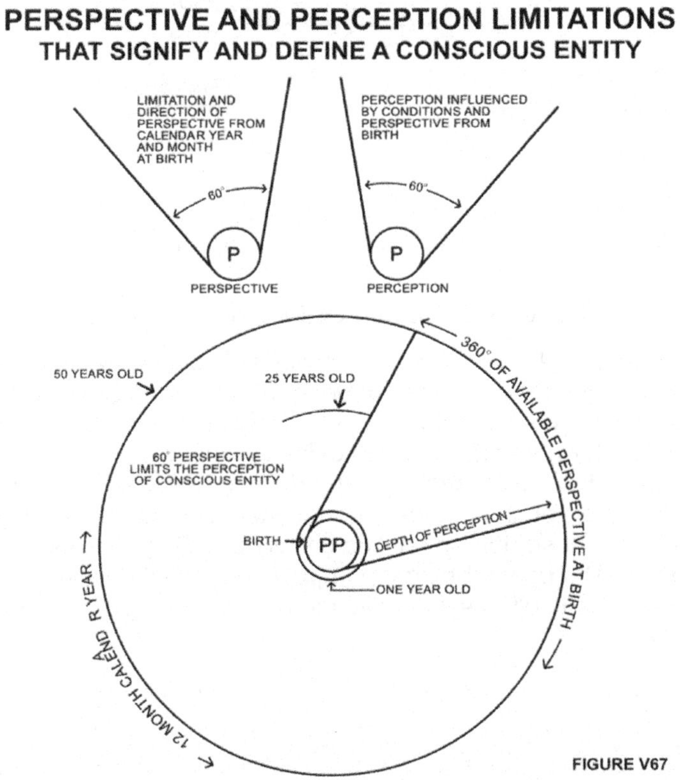

Figure V67: Perspective and perception limitations

In this figure, the direction upon the conscious entity's arrival on Earth sets the perspective of the conscious entity, giving it a specific perspective due to its direction and limited range. The depth of perception is limited by the environment the entity's perspective is subject to, hostility versus civility.

HUMAN COLLECTIVE CONSCIOUS

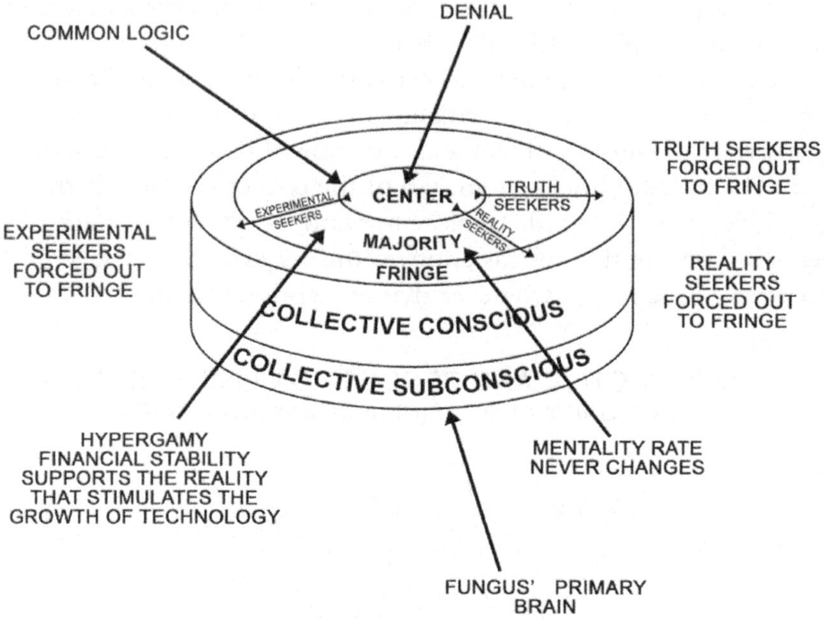

Figure SC1: Human collective conscious.

This figure illustrates how the human collective consciousness functions. It regulates old and new information depending upon the rate of cognitive acceptability within the majority's previously achieved concrete reality. From the Center (denial), made up of lethargic or non-cognition entities. Majority (immediate use) family or reproductive entities. Fringe (conscious denial reminder) artisans, experimental, questioning and truth-seeking entities. These human entities are governed by the collective subconscious. The center being the impact zone of all new information and is Conscious time itself, a systematic method of conscious growth or MCG.

When new information or discovery is acknowledged by the collective, its most important goal is to protect the previously established reality. This is crucial due to the fact that it is under the condition of the Void of Psychosis. The VOP-induced reality is frugal for two reasons. One, it is the only concrete reality it has though always in transition to new information and discovery. Two, that reality needs time and acceptance before acknowledging in order to remain concrete.

A phenomenal example of the above is the O.J. Simpson case. After two decades most Americans now believe he did it. This is a good example of denial until the majority comes to accept a fact because of the collective subconscious. Another example is Vincent van Gogh's artwork. It had been pushed out to the fringe and took 30 years before gaining recognition with the mainstream or majority. Giordano Bruno, a well-established Italian philosopher, was burned at the stake in 1600 for his views in cosmology that stars in the night sky were suns that had other planets orbiting them. Whether the information is experimental, truth or reality, it can be denied for eternity or just simply postponed so it can be used constructively.

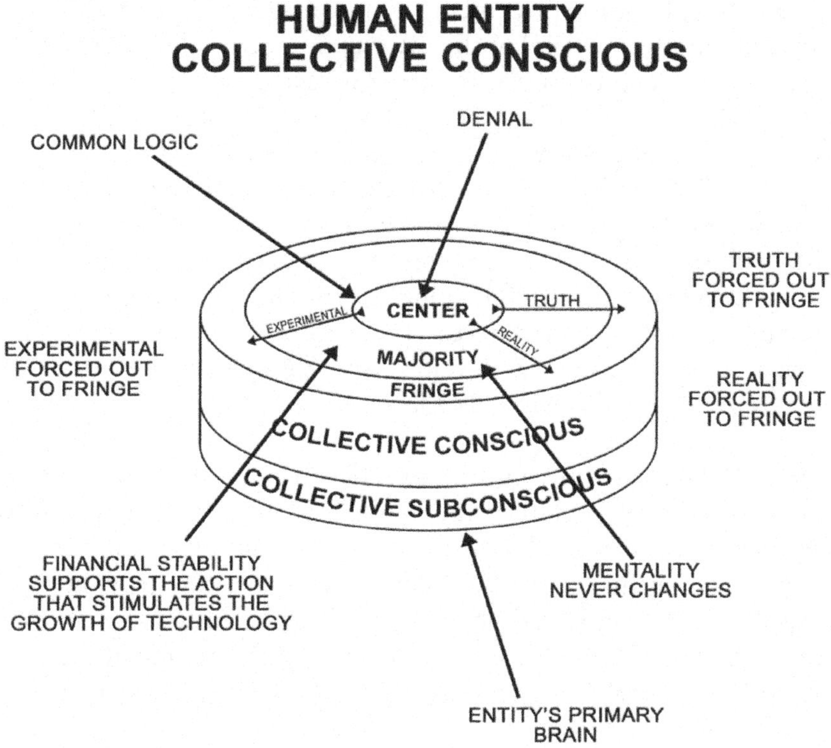

Figure SC2: Human entity collective conscious.

This figure illustrates how a single human entity's consciousness grows. As new discoveries and information enter the brain, they are quickly distributed according to the impact they have on the reality the entity has previously established as concrete. The subconscious protects that entity's reality with this configuration, center (denial), majority (immediate use), or fringe (apply when ready). This protection is an information impact zone and is vital in keeping the reality of the entity safe and sane, and is revealed later when that new information can actually serve a constructive purpose. This system of information regulation is actually Conscious time itself, and cannot increase or decrease without threatening the *fungus's* goal of TES.

The single human entity collective conscious is a miniature version of the human collective conscious (Figure SC2). The distribution of new information and discovery by the collective subconscious protects the previously established concrete reality. If not immediately used, it is placed in temporary holding places – center, majority, and fringe. As bad as denial is in our day-to-day lives, it actually works as a reality buffer, holding back what could shatter a single entity's well-constructed reality even if delusional. Within the mind of this entity is a constant struggle of the placement of information into its need compartment; it is crucial that the information be used when the time for its use can be constructive. The leading cause of human death is not heart attack or cancer; it is suicide and is directly related to psychological conditions with the top being Borderline Personality Disorder (BPD). The distribution of information overload and lack of formulating realities with that information seems to be the cause of such conditions. The entity's subconscious is extremely massive and capable of sorting and filing all input, functioning like a fungal network to sustain the reality growth necessary for the human entity's participation in the technical development that one day will influence the entire human collective conscious.

Every single human entity in the world dealing with this struggle works together to form and maintain the human collective conscious growth. The process of checks and balances is constant among all coexisting entities; together this forms the collective subconscious growth. This is the result of HTO and VOP that form the single entity's PPL, constantly referencing, acknowledging, and verifying in order to make a concrete reality that stimulates the growth of a consciousness to its maximum.

All human entities are basically dealing with a PPL for trauma, spending the rest of their lives trying to accept or deny their childhood trauma. All human entities are basically dealing with a PPL formed by trauma and spend the rest of their lives trying to accept or deny their childhood, whether the entities have suffered abuse or deprivation of any kind. Even if the entity had a peaceful childhood, it may see its society as the abuser to overcome. This very detailed existence is what makes every entity totally different by design, enforcing the growth of the entity and thus the collective. Because this learning by information can only be handled at a reality-forming rate, with each entity all together, this rate is what all time actually is; conscious time renders space time or any other evaluation concept of time irrelevant.

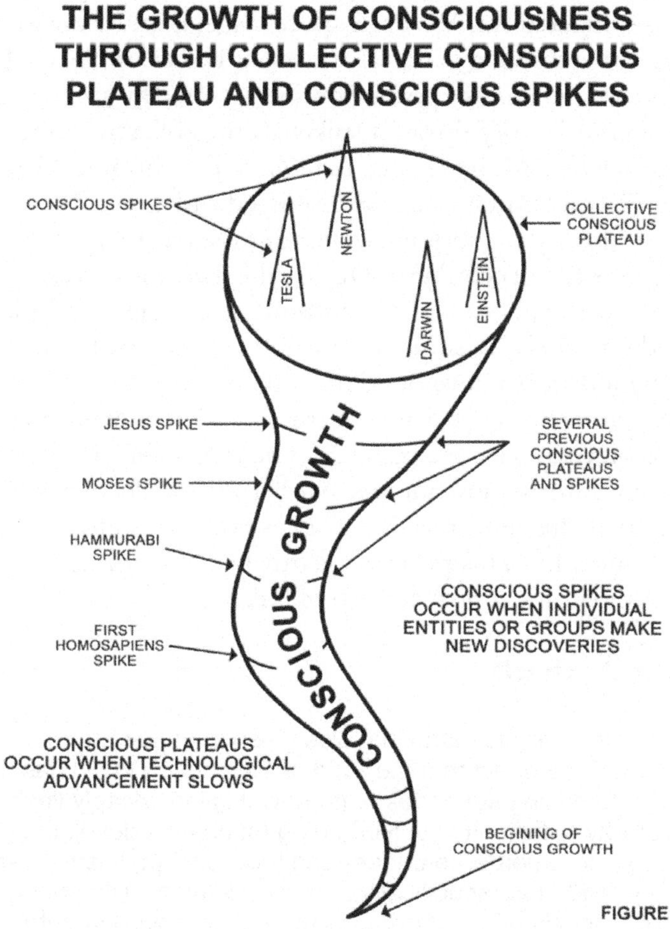

FIGURE T71

Figure T71: Conscious Plateaus and Spikes

This figure is a good visualization of the growth of consciousness (*fungus*) after reaching the spore stage. Within the collective consciousness is the growth of technology, as technology slows its plateaus. This allows for individual spikes within the collective consciousness to rise as new discovery and information that excels the growth of technology again, till it reaches another plateau.

From the beginning of the growth of consciousness to date, we can observe and calculate its growth as a single growth psychology. It may be just a claw, camouflage, the ability to fly, but all forms of evolution and adaptation are technical advancements through thought forms. Through several mass extinctions, plant and animal life has succeeded in this same vein with burrowing, caves, and eventually fire. This shows the necessity of each conscious new thought that makes that very little difference in the need to grow a bigger brain and stand up. After the *fungus* achieves bipedalism, VOP, or spore stage, it is already in the development of technology and information. For example, on Earth information can come in the form of utilizing the new size of the brain. This shows in our history through spikes to the collective in forms of philanthropy such as Moses, Jesus, Lincoln and technology forging of metals, electricity, computer etc. As discussed early, conscious growth at this stage of growth cannot grow without the advancement of technology, thus the (Technium). During the Technium *fungus* phase, the collective conscious adopts the current technology until it is standardized onto itself eventually becoming mundane. Beneath that process is the individual development of discovery and technology that makes spikes of technological advancement and information from the previously achieved collective plateaus. The individual entity spikes feed this new-found information and discovery to the collective; and as the collective applies this new info, it advances conscious growth to a new plateau of mundanity. This process eventually helps the *fungus* prepare for evacuation day and a post-terrestrial life.

The Void of Psychosis

TL;DR: A deeper analysis than what was given in earlier chapters so as to fully demonstrate the effect that the Void of Psychosis (VOP) has, being generated by increasing awareness and how it is inextricably linked to the awareness of any sufficiently evolved being (such as a developing spore) within the universe. In order to understand this concept, it is necessary to define Identity, and more specifically, concrete identity. The concrete identity of a conscious entity pertains to the mostly environmental realities like gravity, aging, reproduction, pain, shock discovery and more that derived from the interaction between the environment and the entity's sensory inputs that give clues to that entity as to the distinction between itself and other outside factors, how that entity relates to those other outside factors (which include both other entities and non-living objects, the changing environment, and its conditions that surround it), and the psychological condition of that entity do to the fact that it does not know why it exist or where it came from originally.

The Description of the Void of Psychosis

For conscious entities that possess the desire to inquire about their origins, the ramifications of cognitive awareness of HTO (Hidden True Origin) imparts a void in the concrete identity of these entities. Once formed, this void manifests as a lack of knowledge about the True Origin of these entities, as well as the ontological nature of that entity or group. This information or none information inevitably causes a psychosis or void in the concrete identity. That entity then is forced to attempt to fill the void through various actions that help further signify its concrete identity. This can occur at an individual or group level. It is this dynamic and EOS which is at the nucleus of the particular types of action necessary to generate technology for achieving TES (Technological Evacuation Standards).

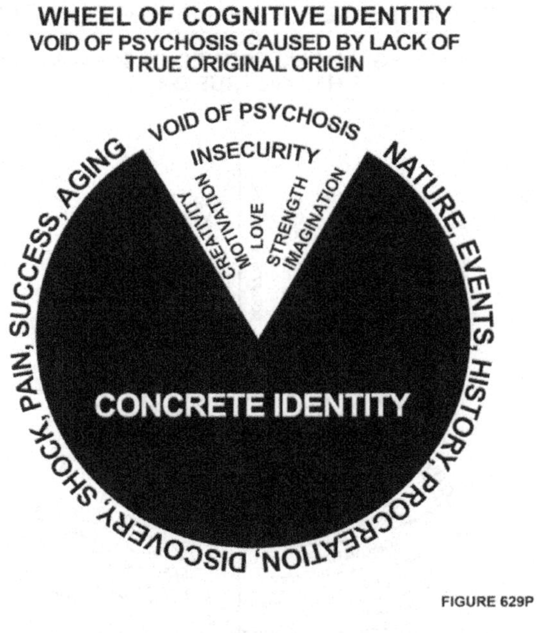

Figure 629P: Wheel of cognitive identity

One's identity can be imagined as forming a nearly complete circle which contains concrete elements that work together to form a perception of how one fits into the world and society. However, a percentage is missing due to the fact that the true origin of all life is unknown. The outer circumference of the wheel shows elements that establish a stable or concrete identity; yet the missing space (void) resembles the unknown source of consciousness and true identity, which is linked to true origin.

To continue to grow consciousness, whether that be at the level of a collective or an individual, that consciousness must be vexed by a state or condition that brings about cognitive dissonance. In general, the largest factor that brings about these states is the lack of knowledge of true origin. This instills severe insecurity in individual conscious entities as well as in the collective consciousness of the *fungus*. Through the actions of the individual, as well as the collective, attempts are made to signify a more concrete identity, or, in some rare few, the attempt to discover true origin is an alluring enigma that is persistently pursued. This is an attempt to fill the void. This aids in finding meaning and purpose in life. In doing so, the psychosis is reduced incrementally but never fully resolved. This advances the growth of the *fungus* as it approaches the formation of a spore.

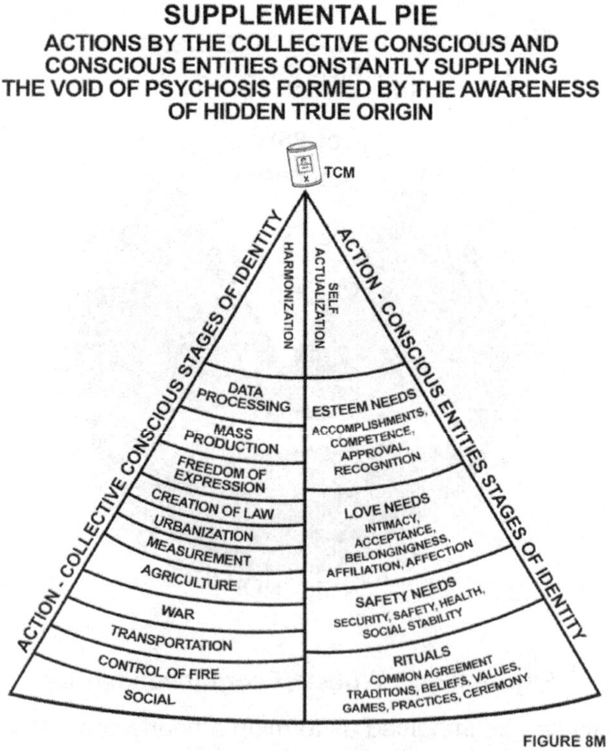

Figure 8M (Supplemental Pie)

This figure is a representation of the various ways in which an individual or collective consciousness will attempt to fill the void. Each layer in the rotated image (eye at the top) describes the perspectives that each "eye", or level of awareness, is concerned with, whether it is cognizant of these perspectives or not, on the way toward evolving to a TCM.

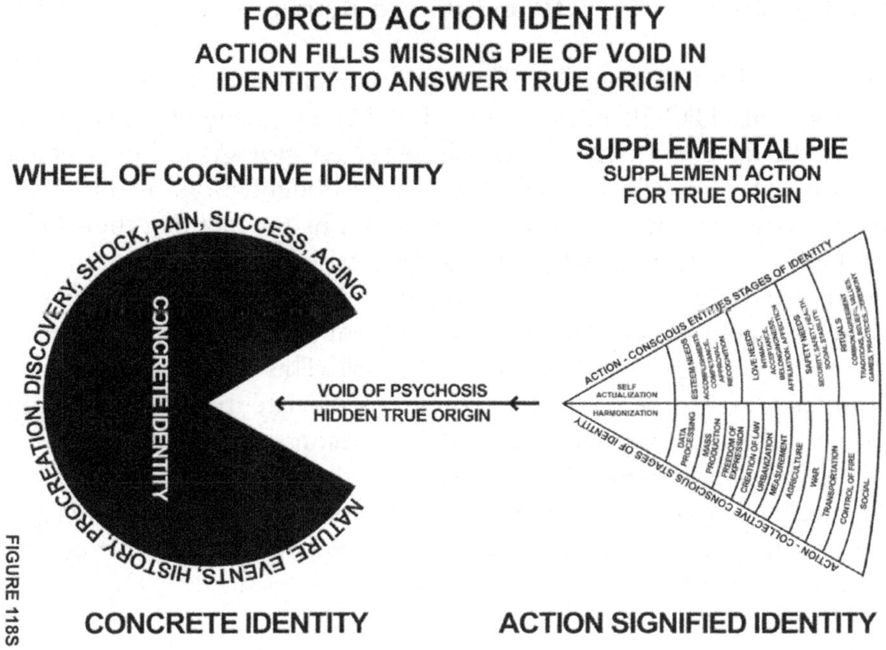

Figure 118S (Forced Action)

This figure shows how the various needs and concerns of an individual attempting to form an identity in Figure 8M fit into the VOP in Figure 629P.

Attempting to fill the VOP with these actions is the process by which identity is solidified in the face of HTO.

In order for the universe to generate MCG, a state of growing awareness in a developing fungal growth enforces conditions that are required to develop psychology in that fungal growth. This developing psychology eventually reaches a point in which a growing awareness reveals that information is fundamentally hidden from that psychology. In coming to this realization, conscious entities within that fungal growth are forced to act through the development of a psychosis that pushes these conscious entities toward specific actions that promote a more signified identity. These actions ultimately promote the development of tools and technologies that allow the fungal growth to survive the end of its birthplace. In so doing, the conscious entities of the fungal growth become a spore of that growth and exist as a TCM.

The Dynamics of the VOP and Forced Action

Since every BCT is brought into the universe under the forced condition of HTO, there exists a void in their complete identity, which causes the development of psychosis in the psychology of the collective or entity. Here, psychosis is the discomfort brought about by this lack of a solidified identity. This is brought about by the fact that they do not know who they are, why they are here, where they came from, where they are going, and, ultimately, what the significance of their existence might be. Ultimately, a lack of identity forces motivation to act in order to signify a more identifiable self. This new, more identifiable self is the continued attempt to alleviate the psychosis and fill the VOP. The very act (all actions) of creating a more identifiable self contributes to achieving TES. Without these dynamics, which are the mechanistic process by which the universe promotes conscious-growth, the potential of a Big Bang that results in a dud (no formation of a *fungus* that will achieve MCG) would be too great.

Conscious Time

What is being conscious and what is time? Let's first start off with being conscious as a growth process and not a static state as it is commonly referred to. Conscious growth is a brain receiving information and experience resulting in discovery. This is true in the single conscious entity and the collective conscious focusing on humans. Consciously growing means that each discovery made by humans is a new layer of reality on top of all the discoveries previously made in human culture. Looking back thousands of years, this very slow accumulation of discoveries has led to the reality you and I came to make today. This growth process in and of itself is fungal like in the billions of tidbits and fragmentation of awareness that feed the hungry mouth of human curiosity. If looked at in this way, there is also a distinct amount of information that an entity's brain can receive without destroying or at least disrupting their reality. The same goes for the collective as it can only identify with itself as one reality at a time. In fact, if the reality of the collective takes a big jump, it can cause pandemonium, social suicide, or collapse. Most nations go to war because the other advanced nations are reflective; this causes the suffering nation the ability to identify with its secure reality (tradition) causing it to kill for its own survival. This is why conscious growth has a strict timing to its advancement. A good example is the information in this book *Fungus*

Theory. It may be too much too fast and has information that cannot yet be applied by the entity reading it. Therefore the collective has a reality buffer called denial that allows the recipient to take time to formulate new realities. Galileo is another example of too much reality-changing information too fast that can disrupt the reality that the society needs to function in the development of the technologies of the times. He was put on house arrest for the remaining 10 years of his life. This made a reality buffer and eventually all his discoveries were accepted when the information was no longer deemed a danger. This makes denial a very important element for conscious growth. Lagging information is also a danger to the *fungus* growth toward TES. Language barriers and travel and transportation of goods are a danger making a nation lag when progress is crucially important for the survival of humans. This could be a good reason why space aliens do not intermingle with our society as this would tragically disrupt millions of years of stable reality building for us humans, and the aliens know this. It will probably be that we discover aliens on our own long before we are able to adjust to the reality-changing information that could occur. The *fungus* has a strict balance between information overload and starvation controlling the reality needed to push technology to TES.

Conscious time is regulated biologically by the rate at which information and experience intersect the human brain. Mitosis time is the clock in which conscious growth time grows. Whether it is a conscious entity or the collective, it is strictly governed by the amount or lack of information and experience can turn into discovery without disrupting reality. This makes conscious growth time the real time in order to form a past, present, and future, the real essence of a *fungus* formulating and innovating technology in order to meet TES.

Single Entity Conscious Timing

For the billions of entities needed to conclude new discoveries for a collective to excel in technology, each entity needs to focus strongly on action needed to take place in order to discover and submit to the collective. There is a strict rate at which this information can enter and still be functional for the collective. As shown in figure CS1, you can see a growth structure of information in the restriction of truth or reality. This is a rate or governor to the collective that solidifies discovery as absolute and something the collective uses to catapult into new discovery. If the growth of reality stalls or slows from the lack of information and discovery, it may become distorted rendering future technological growth unreliable or doomed. In a sense, all technology must be completely earned by the human collective to assure post-Earth existence.

In order for a single entity to commit an action that results in discovery, it needs to identify with the reality of its own cognition and environment. This is the foundation for a dopamine secretion that encourages and inspires such an action. Reality is determined by the entity as a concrete and absolute truth in whatever particular timeline in the conscious growth. This also makes reality only valid in the time in which it has been made, establishing the innovation of technology during that time period. Because technology is the *fungus's* primary by-product, this is what establishes the reality at the moment of discovery of conscious growth. For instance, Arthur C. Clark thought there were trees on Mars because the telescopes at the time blurred images, making it look like trees and shadows. After the improvement of the telescope, it was discovered there is no life there, changing his reality.

The entity itself is also only cognizant of the elements in its environment that enforce the action needed for technological growth. Example: if you could watch your fingernails grow, you would be more conscious of the time you have left to live. That being your new reality would change your goal to be more of your own survival and not the collective's primary goal (action for innovations). Even watching your hair grow in the mirror would cause you to act against the *fungus's* primary goal of TES. It is essential that an entity forget that they are in evolution, coexisting with other humans in their constant evolution. Conscious time is a rate that humans cannot consciously apply to their simultaneous relationship with aging and time.

Each entity's subconscious understands the *fungus's* primary goal for TES and therefore does not spend its life reinventing the wheel. Mitosis time is the governor of conscious growth and is the most effective time signature, making conscious time the only time that counts in the entire universe. As long as humans are biological, this time governor should never change. If humans become mechanical in any way, there is an absolute failure rate with mechanics but not with biologics keeping mitosis time, the true time.

Collective Conscious Timing

Much like a single entity, the collective is a single animal in the same way in the human sense. In order for the collective to develop the technology for TES, it must grow its technology at a trusted rate so it can be relied on for eternity. Because innovation and technology are layers of discovery, it is essential that each one is achieved in an almost sequential order. For instance, you can't make a microwave oven before a toaster and you can't make a toaster before the advent of electricity. James Burke from *Connections* has made the best example of this strict layered process of innovation that totally relies on one innovation making the next one possible. As a system, the *fungus* uses humanity to fund the tech growth that will save the *fungus* from annihilation. A very interlocking discovery process that relies on the post-Earth and ESS – once it leaves Earth there it's never coming back.

Collective Subconscious Timing

The subconscious is the *fungus's* brain. It seems to know its function and purpose in the design of the universe. It behaves as though there is a strong intention driving it while keeping something a secret. It's as though it knows it is tricking humans into action and holding back the reason. Confident in its pursuit, its momentum has never wavered or slowed down in any way that we can see since recorded time. It is also the profound evidence of HTO by guinea pigging and literally using humans for its ultimate goal. Like a pharmacy built inside each human, the dopamine reward mechanism makes sure that there is an endless need to discover. The subconscious is also a governor of the amount of discovery and information that the collective conscious can handle at once. This is EOS that the growth of a consciousness is the universe's intended goal and the end of Earth is EOS for the growth of that consciousness to its maximum, or MCG.

This slow discovery rate from humans falls into the abyss of the subconscious collective; this helps it push forward to the next things that the conscious needs to become conscious of and no more than that. It is directly connected with each entity's subconscious that drives the conscious to act, always taking form the previous discovery and self-giving the impulses necessary for the next discovery. This order is critical for the growth of technology and is also evident throughout history, showing that it is totally governed by the collective subconscious. By using dopamine to power the human ego, it is thought we are its puppets. Its intentions show from the very beginning about 2 billion YA, first formation of plant and animal fungus (PAAF). AI and other information input to the Timing of conscious growth still fall under the same method of reality forming within the collective. The split from plant and then animal is the setup and intention for the growth of consciousness as though it knows it needs a human to save it from the end of its birth place, Earth.

ITM As early hominids began to evolve on Earth, many of the underpinnings required for emotional intelligence had already been evolving for quite some time. "Reptilian" emotions associated with the amygdala relate to hate, whereas in more evolved animals the amygdala, in conjunction with other brain centers, is associated with love. Animals, representing more advanced forms of life than reptiles and other prior aquatic organisms, have more developed brains and brain regions for emotional response. Thus, hominids had most of the hardware for evolved forms of emotion to develop but hadn't yet developed the advanced circuitry to reinforce these modalities of behavior.

As discussed prior, energy efficiency is pivotal for the development of the growing conscious brain. The move out of water and onto the land in order to respire in free air was discussed as being a pivotal evolutionary change in order of increasing energy transport through respiration, as was the condensing of nutrients into vegetables for land mammals. Similarly, the discovery of the many uses of fire, chiefly its implementation in cooking, had a transformative effect on early BCTs.

"In their article, the authors note that the change from raw to cooked food had a profound impact on human development and behavior. As a result of cooking food, the body uses less energy to digest and break down food, giving other systems more time to develop. During this process, the structure of the human jaw and skull is also altered. The change allowed humans to develop novel social and behavioral systems without having to spend time foraging and digesting raw food on a daily basis. In some scientists' eyes, fish consumption marked a quantum leap in human development."[106]

Given that these BCTs were able to consistently intake foods that were more easily broken down and absorbed, the energy conserved afforded the biology of these early BCTs the chance to continue growing centers of the brain and other neural circuitry. This increase in complexity is likely the catalyst of the development of complex neural feedback systems that allow for the development of a psychology over time. With continued evolution of this psychology, the VOP was sure to appear.

As BCTs continued to evolve, their development of tools as well as the implementation of these tools using their more dextrous appendages (opposable thumbs) and increased cognitive abilities allowed for greater neural development. These primitive tools eventually led to the ability to support greater numbers in groups/cliques and promoted the early social behaviors that are associated with the more complex animal emotions found in humans.

"Sometime in the Stone Age, human artists began experimenting with a new form of visual art: drawing. Now, from the ancient rubble that accumulated on the floor of a South African cave comes the earliest-known example — an abstract, crayon-on-stone piece created about 73,000 years ago."[407]

This demonstrates that humans began to think about leaving forms of messages (whether inspirational or informational) to others. This act also demonstrates that a VOP had formed by this time due to the fact that thinking about leaving information for others to see at later periods of time is occurring, as well as thinking in abstract ways. Abstractions require a level of psychological inquiry and projection that is relatable to all…in essence to their current condition of living. Leaving these marks or symbols that others can see and now relate to the condition of living itself leads to the very beginnings of the questioning of true origin.

"Many of the hand stencils appear in small recesses of the cave that are hard to reach, suggesting the person who made them had to prepare pigment and light before venturing into the cave to find the desired spot. The markings themselves are also interesting because they demonstrate symbolic thinking."[408]

These drawings are a sure sign that humans began pondering their existence, what it meant to be human, and how to communicate information to other groups of people, whether in the present or the future. However, further development of the VOP would invoke questions of origin and a desire to relate to the creator of one's being. In this sense, the idea of appeasing the creator or creators comes into the scope of human consciousness. As humans are a limited and finite organism, they would not be capable of truly understanding the wishes and desires of such an omnipotent being. In order to appease these beings, offerings and sacrifices represent a reflection of the internal workings of the psychology of those making the offerings. In essence, when humans began to make offerings and sacrifices around three to five thousand years ago, there is profound sociological proof that the effects of a firmly developed VOP are forcing actions in these more modern humans.

IPT Life on Earth has changed dramatically since the earliest humans walked across land and established small communities. However, many of the needs and concerns of humans today remain the same. What has changed are the ways in which they are fulfilled, with some exceptions. For example, the average human of today is still very much under the influence of the needs within the first eye, as described in Figure 8M. In fact, most of the time spent by humans is spent ensuring that the needs of the first eye are met. This encompasses their need for shelter and protection, which is delivered to them by their parents, or by obtaining employment and paying for housing if they are independent adults. However, in the past, the need for shelter or protection was not guaranteed by parents but was something that was fought for and provided by nature. While some of the concerns in the second eye share overlap with those in the first, there is a more psychological nature to this eye in that it doesn't consist of only the necessities for survival, but includes concerns for mental well-being. Maintaining optimal health, gaining employment, being surrounded by individuals (not being homeless or a loner), having possessions, and other forms of integrating with society allow for a more grounded and stable psychology, and are essential for the success of a bipedal organism. The third eye drills a layer deeper and is a bridge between animalistic survival instincts and higher-order psychological desires that stem from the cognitive awareness of one's place in the world, the value of one's time and energy, as well as understanding the importance of interpersonal relationships. These desires and concerns deal more with the emotions of love, fulfillment, belongingness, and joy. Since this eye includes the formation of intimate relationships, it demonstrates an awareness of the desires and uniqueness of other individuals. It also includes the desire to give affection, which could be displayed in friendships and intimate relationships alike. The fourth eye is focused on the quality of the self-perception of an identity. This means that there is now awareness of the fact that simply meeting the physical needs and psychological desires in prior eyes is not all that one can strive for. In this eye, personal achievements and the esteem brought about by targeting and achieving greater goals take up a significant portion of the time and energy of an identity operating at this level. However, this individual will likely not sacrifice their own status, financial security, or time due to being concerned with their connections and relationships. Some examples of individuals at this level would be a parent who holds a steady job and has a family but has always wanted to express their individuality through art and music. This type of pursuit will likely not require large sacrifices but it does stand as a personal achievement in which they can take pride. However, someone in the fifth eye will try to realize their full potential, no matter the cost. This is because this eye is concerned with self-actualization. Individuals in this level often sacrifice jobs, relationships, money, most of their time, and even their lives/health in pursuit of what they believe the world (*fungus*) needs, which is what is ultimately required to achieve TES as a TCM.

BP Earth's *fungus* is continually striving to reach higher states of conscious awareness through the actions that are forced by the needs and concerns in the different layers (or eyes) of Figure 8M. These forced actions are the main drivers of evolution in the *fungus* so that TES will be achieved and a TCM can be formed. A TCM, as well as the point at which TES is achieved is the sixth layer in Figure 8M. At this level of conscious awareness, the *fungus* is merely concerned with achieving ESS so that it may survive indefinitely in the cosmos. Achieving ESS requires more than the act of becoming a TCM. It requires that the TCM begins developing technologies and systems that can sustain it no matter where it exists in the universe. Thus, ESS is achieved by the fortification of the TCM. This requires time and experience in the cosmos to determine exactly how to meet these needs, which is the same as what is required to reach TES on a biosphere. This matured consciousness, existing in a MCS, will go on to promote further conscious-growth in the cosmos through its various interactions and potential encounters with other fungal growths.

BBP Ultimately, Figure 8M lays out the resulting stages in conscious-growth due to the conditions that are put in place by the fundamental limitations of the universe, the main contributor being HTO. This is occurring in parallel throughout the cosmos, resulting in the forced evolution of any life that may form on a biosphere. Each of these conscious-growths may have slightly different conditions on their respective biospheres, but the general pattern of needs follows that in Figure 8M. The result of this forced action on each conscious-growth throughout the universe demonstrates that the universe is a plantation for the growth of consciousness and that its purpose is to push consciousness to a defining limit. The fact that every star goes nova, that every black hole will eventually devour its surrounding stars, and that the universe will continue to expand indefinitely are all demonstrations that no space is safe indefinitely and further evolution and increasing intelligence will always be required to surmount each of these hardships. Hence, the universe spawns consciousness and forces it to evolve to the point of ESS. Eventually, these consciousnesses may meet and share knowledge and resources for the UGCH.

EOS

- HTO, the basis for psychosis
- Inevitable death, the absolute motivator for spore stage consciousness
- Action, compensation for the lack of true identity
- Manipulation of environment
- End of Earth preparations

Potential Evolution of Psychology in Organisms Biospheric Economy

Psychology and Economy: Bidirectional Feedback Systems

TL;DR: Each biosphere throughout the universe is set up with a very strict economy of energy supplied only by its host star. A biosphere is the embryonic stage of a fungal growth that is strictly governed by the slow and gradual self-feeding growth process called starvation. It ignites the predator/prey (cannibalistic) economics that cultivate the *fungus* to its bipedal stage or spore stage. When the *fungus* develops the brain capability to supersede the predator/prey economics, it must turn to a more post-terrestrial form of food energy in the new environment it will migrate to, leaving its birthplace economics to die much like the placenta and sac and umbilical source a human baby is born from.

Each planet (biosphere) that is home to some form of fungal growth does so under the strict limitations set in place by that planet's supply of energy, raw materials, and the time with which it remains hospitable for life to cultivate and thrive (its matter/pure-energy economy). In Earth's case, within 250 million years the oceans will evaporate and make it inhospitable for virtually any form of life to exist. These economic limitations coupled with a limited timeline are the key motivators in the evolution of Earth's fungal growth and operate via predator-prey dynamics, which is an aspect of the second fundamental limitation of the universe: the threat of inevitable death. Given that these conditions or absolutes regarding the finite amount of resources play into the way in which organisms behave, there is a clear link between the Earth's matter/pure-energy economy and the psychology of developing bipedal organisms. Since this link exists, it is also equally valid to suggest that an influence in the opposite direction can occur, namely the psychology of an organism that possesses a VOP is capable of influencing the matter/pure-energy economy.

ITM Early hominids were at the whim of the environment for the most part. Tool making had not become widespread and large cities and towns were not yet something the BCT was capable of creating. Early hominids and early humans would exist in smaller groups for a number of reasons. Larger group organization takes more complex brain development, the ability to sustain the dietary requirements of each and every member of the group, and a way to clearly define roles or to adapt to changing roles. Given the complexity of such tasks, early BCT development occurred in these smaller groups for some time.

"Indirect approaches of group size, using occupational or sleeping surface areas and ethnographic analogies (e.g., refs. 3, 6, and 7), suggest that Neandertal groups were small, from 10 to 30 individuals on average per site (e.g., ref. 3)."[109]

In these smaller groups, the matter/pure-energy economy of the Earth plays a significant role in affecting the psychology and behavior of the collective. The collective is acutely aware of their needs. However, they are at the whim of what is available and provided by the local environment. These early hunter-gatherers would go out and hunt for food and potentially steal from other groups of hunter-gatherers if the opportunity presented itself. In this way, they were totally influenced by their surrounding environment. Their behaviors and psychological development were intimately tied to what nature provided.

With time and experience being hunter-gatherers, humans began to incrementally develop tools and started experimenting with different ideas of resource gathering. This would come in the form of harvesting crops in order to produce sustainable sources of food.

"A new discovery by an international collaboration of researchers from Tel Aviv University, Harvard University, Bar-Ilan University, and the University of Haifa offers the first evidence that trial plant cultivation began far earlier -- some 23,000 years ago."[110]

It is the advent of farming and agriculture that would begin a slow and persistent change to the way the BCT began interacting with the matter/pure-energy economy. As agriculture became more refined, and irrigation systems were invented, early humans began living in larger and larger groups. Eventually, cities and towns began significantly expanding in size. With this growth in size, more evolution occurred in the brains of BCTs to account for changes in social structure. It wasn't until the advent of the plow that truly massive societies were capable of forming and lasting.

"While historians aren't entirely certain of where the plow originated, evidence suggests that the Egyptians and Sumerians were among the first societies to employ its use around 4000 B.C. ...That all changed in 2000 B.C., when the Egyptians first hooked their plows to oxen."[111]

As the ox-driven plow began to transform civilization, it was liberated from the traditional matter/pure-energy economy being the direct influencer of psychology. Over thousands of years, trade routes over land and sea, as well as the introduction of various plant and animal species to areas where they were not indigenous, allowed the psychology of humans to affect the economy. For example, early dairy farming significantly changed local economies and also the evolution of humans themselves!

"Genetic studies show that goats and other livestock accompanied the westward spread of agriculture into Europe, helping to revolutionize Stone Age society...Prior to the arrival of domestic cattle in Europe, prehistoric populations weren't able to stomach raw cow milk. But at some point during the spread of farming into southeastern Europe, a mutation occurred for lactose tolerance that increased in frequency through natural selection thanks to the nourishing benefits of milk."[112]

The agricultural ability of these humans allowed them to transform both the economy around them as well as themselves. This trend would continue as humans continually evolved until the modern day.

IPT In modern times, we see that humanity is influenced by economic factors, such as drought or reduced supplies of various crops and goods, as well as is capable of changing economies at a large scale. Mass hysteria can be seen in a variety of situations where humans face floods, war, famine, or other unfavorable conditions. Every one of us today has seen food fly off the shelves and common goods sell out when hard times are predicted on the news. This type of economic influence on human psychology is still very strongly rooted in the dynamics of the *fungus*.

What is novel is humanity's profound influence on the matter/pure-energy economy in recent times. Humanity's recent societal activities are responsible for changes within Earth's conditions, such as the cleanliness of the air or the level of heavy metal that circulate through the air/water, which vastly affects all life on Earth. By warming and cooling various regions, the most obvious effect is the extinction of various plant and animal species that became accustomed to thriving within certain temperature ranges. In fact, scientists are speculating that humanity has now entered the sixth mass extinction.

"Unlike previous extinction events caused by natural phenomena, the sixth mass extinction is driven by human activity, primarily (though not limited to) the unsustainable use of land, water and energy use..."[413]

It is at this stage that the developing spore is capable of feats that were unimaginable at prior points in the development of the *fungus*. This ability of human psychology to eventually influence the matter/pure-energy economy is pivotal for broadening awareness and generating concern to fuel the advancement of innovative technologies.

BP As Earth's *fungus* evolved, increasingly complex conscious organisms began to develop. When sufficient complexity in consciousness generated an organism that possessed awareness, psychology began to develop. With sufficient time, increasing awareness generates concern within an organism and this concern begins to develop a psychology that possesses a VOP. It is this process of producing a psychology that possesses a VOP that allows the possibility of a bidirectional influence between economy and psychology. Prior to the existence of psychology, organisms were only capable of exhibiting change to the economy as a matter of happenstance and normal biological function. Algae may have transformed Earth's atmosphere, though this was not done out of any concern or desire to make changes; it simply occurred because this was part of the natural function these organisms evolved to perform. With the appearance of BCTs possessing a psychology that contains a VOP, activities outside normal biological function are associated with transformative activity in regard to the economy. This is a necessity for the *fungus* to achieve TES and to develop technologies that will allow the spore to exist in ESS as a TCM.

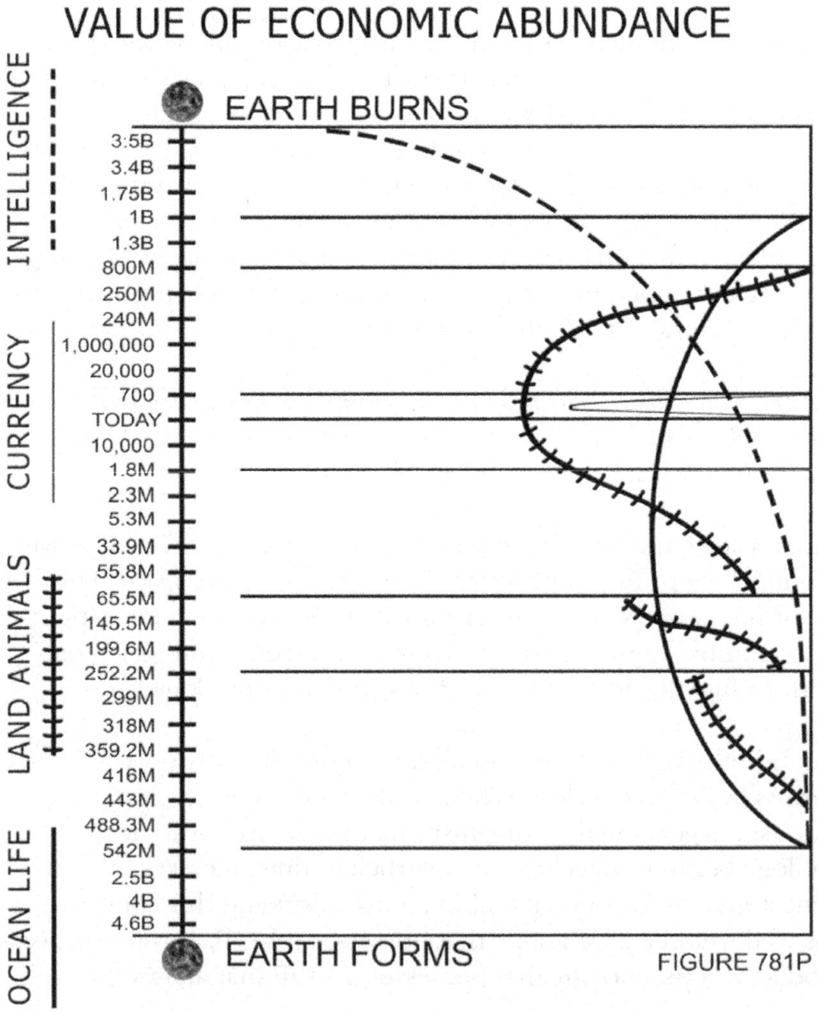

FIGURE 781P

Figure 781P: Value of Earth's Matter/Pure-Energy Economy

This figure displays the various outputs (past, present, and future) of Earth's matter/pure-energy economy. The economic cycling of raw material and energy resources is shown with an overlay of intelligent growth during these time periods.

Note that the increase in intelligence is an exponential curve but really begins to take off once the BCT (with its VOP) begins innovating and transforming societal structures.

In essence, humanity needs to begin transforming (and inevitably harming) its environment to become aware of the importance of its environment, which will generate concern. This then allows the development of more efficient and sustainable technologies in order to survive both on and off the Earth. By the time Earth is no longer habitable, all of the various interactions between the matter/pure-energy economy and the psychology of BCTs will have allowed for the formation of the spore of the *fungus*, and the post-terrestrial future of the *fungus* is ensured.

BBP Across the universe, fungal growths will take root and begin to evolve on their respective biospheres. Each of these growths will begin to evolve in a manner as described previously. After the assemblage of the most primitive life, which possesses minimal consciousness, their consciousness begins to expand through the evolution that takes place in their respective environments. Without a VOP, there is little-to-no reason to develop technologies as there is no desire to alleviate the void through various actions. Thus, the VOP is a pivotal development in any fungal growth that catalyzes technological development and shifts the behavior of BCTs developed by these growths to be more concerned and forward-thinking. It is the attempt to continually fill that VOP that ensures the development of a spore.

Each fungal growth on extraterrestrial biospheres will develop its own spore, which may go on to encounter others in the cosmos. This possibility allows for further attempts to rectify HTO and alleviate the VOP. What one spore may have learned along the way toward achieving ESS as a TCM may differ from another. Thus, the sharing of information between differing spores promotes MCG. With enough time, each of these spores will be pushed further toward MCG in preparation for the UGCH.

EOS

- Cannibalistic self-feeding processes
- Evolutionary growth of plants and animals
- Religion
- Technology
- Emotional reaction
- Behavior
- Forced growth of consciousness

The Money Economy on Earth

TL;DR: In order for technology (the final kingdom) to grow and be successful with TES, it must become civil in order for the pontification of invention and innovation to spawn. For this to happen, the processes regulating labor and goods must be controlled by a highly intelligent exchange of value for it. Until the *fungus* is at its permeated collective doom acknowledgment (evacuation day), a monetary form of reward is indeed needed for the growth of technology. This process subconsciously prepares the collective for its permeated doom acknowledgment, perhaps dissolving the money economy and resorting to a TCM motivated economy much the same way the Egyptians did for the Farrow.

ITM On Earth, the psychology of the *fungus* was such that transactions took place but were non-denominated exchanges. These exchanges occurred between microorganisms, animals, and the environment, and were principally involved with the energy delivered from the sun, as well as the various raw materials that existed on Earth at any given time. This was the evolutionary status quo throughout the evolution of life on Earth. This system contained inherent inefficiencies, like the inefficiencies in the systems of early single-celled organisms that had been improved upon throughout evolution. In fact, it is these organic inefficiencies that are a driving force in the evolution toward a more perfected terrestrial organism: a BCT.

When the *fungus* begins developing a spore, where it produced a BCT that possessed a complex enough consciousness to house intricate emotions and psychological needs and concerns, these inefficiencies were addressed in a manner that was unseen up until that moment in time. This novel innovation was the development of a unit of currency instead of trading resources directly. In China, the first currency had been used but ultimately did not revolutionize bartering and trade, which would occur later in Mesopotamia.

"Beginning about 5000 BC, cowrie shells were the first form of currency used in China."[114]

These shells were used as a form of payment in China and represented status and wealth as well. Some time later, in Ancient Mesopotamia, a system based on weights was used to standardize trade, which made these encounters far more efficient and secure as the value of goods could be tied to a known unchanging weight or value. This is important because it alleviates the concerns brought about by the psychology possessed by the peoples of that time. Oftentimes traders and merchants would have to carry all of their goods with them in order to make trades, which was cumbersome. In addition, the security of knowing you are getting exactly what you are paying for outperforms trades based on good for good, as the weights of livestock or other goods can be variable.

The area in which this innovation occurred is known as the "Fertile Crescent" where much of our evolved society has its roots. The larger and more intricate cities in this region required transactions to occur at higher volumes and, naturally, evolution in the way transactions were carried out took place.

"Merchants of the Bronze Age faced the same problem as merchants from London to Lisbon today: how to know you're getting what you pay for in a transaction…Standard weights—used by merchants to trade goods of equivalent value—were invented in Egypt or Mesopotamia 5000 years ago."[115]

The creation of a standardized system by humans eliminated many of the inefficiencies in simply trading livestock or produce, which were inherent in the barter/trade systems that had existed for thousands of years prior. Humans began to relate the value of raw materials in the global economic supply based on another commodity for which a value could be agreed upon by all.

Trade occurred as a transaction for goods, services, and animals. However, human life could also be traded. Slaves were frequently sold in barter or trade in order to provide laborers or workers to take care of property and daily tasks. This is the case because labor, specifically human labor, is a highly valuable commodity. In fact, it is likely the most valuable commodity, though it may not always have been compensated for in a manner commensurate with its true value as seen by the use of slave labor.

With the advent of the aforementioned standardized measurements of weight for transactions and the appearance of marketplaces, the way in which humans began to interact in transactions became more civil. As a result, and again with much time, the way people treated each other became more civil as well. This would appear in the form of paying people for their time as laborers. This is an example of demonstrating the value of human labor as it became obvious and necessary that these laborers needed to be treated better, and as such, compensated. This demonstrates concern for the physical and psychological well-being of humans in general. Thus, the developing money economy also had its hand in influencing the psychology of humans.

These changes in social economics represented a large leap in the mentality of humans, in general. The money-based economy would continue to evolve and transform as various trade networks and a greater variety of goods & services were developed. It is these early advances in how to deal with transactions and how to value human labor that leads to the civilization that we see today. In essence, it is not a large leap to conclude that this developing system of economic transactions helped reduce activities such as rape, pillage, and plunder that are indicative of less civil societies. These changes also allowed for the accelerated pace at which technology began to develop.

IPT In more modern times, these advances in currency and economy have led up to the global economic situation that exists today. The entire world is engaged in importing and exporting goods and providing services to different countries. Some countries are known for their ability to produce the best microchips (as with Taiwan), whereas other countries are known for their cheap export of textiles (as is the case in China, Germany, Bangladesh, and Vietnam). These economies have sprawled out from the very humble beginnings that started in ancient Mesopotamia. These economic networks have grown in a fungal manner, branching and interconnecting once-separated cultures and regions of the world. It is this interconnectivity that has been pivotal to the rapid acceleration of technological advancement. By interconnecting many different cultures, the psychology of humanity has also evolved to adapt to include more tolerance for other cultures as well as more awareness and concern about what is going on around the globe.

It is easy to see just how important the global economy has become in terms of technological advancement; one need only look at the strain on the global supply chain due to the COVID-19 pandemic as well as other impedances to supply chains around the world. Given the enormous scale of various global lockdowns and stay-at-home orders from various governments, the demand for goods and services dropped significantly. Interestingly, the demand for smartphones, laptops, and computers increased during this time as people were staying home and engaging in various forms of entertainment both more frequently and in greater numbers. Interestingly, even though the demand for tech was extremely high, and companies such as Apple made record profits, the growth of technology was still impeded by the friction caused by supply chain issues in other goods. The increased cost of oil, as well as many other goods, was an added source of friction (mainly monetary) toward the development and production of technologies. Due to changes in pricing, certain projects were no longer viable or would not be attractive to consumers at their new projected market price. This demonstrates that technology is highly reliant on a thriving economy. Money, being the motivator of the economy, is the fuel that the economy burns in order to produce technological growth.

The most recent kind of economic transformation comes in the form of purely digital currency that is decentralized and cryptographic in nature. These currencies are part of a class of currencies called cryptocurrencies. The original and most famous cryptocurrency is Bitcoin. Its initial attraction stemmed from the fact that transactions were private due to their cryptographic nature, meaning governments or other individuals could not know who was transacting with whom unless the ID of the digital wallet where funds were being spent was known. This type of technology was created due to the growth in the psychology of humans in one main aspect, which was the fact that people felt that the government did not need to know about every transaction in which they participated. While for the most part true, and Bitcoin did provide a large amount of anonymity, it does turn out that patterns in activity on the network and size of wallets can reduce the level of anonymity of the sender, which has spawned other variations of Bitcoin that help further obfuscate transactional information. Other cryptocurrencies, like Ethereum, have allowed code to be run on the Ethereum blockchain (the cryptographic ledger that stores all transactional information that occurs on the network). This expands the use cases of cryptocurrencies in innumerable ways.

Cryptocurrencies have exploded in popularity since the advent of Bitcoin and this currency has outperformed all other assets in its continual year-over-year growth. This demonstrates that the economy of Earth is once again ripe for another evolutionary leap that will transform the way society functions and will promote the technological advancement necessary for TES. Needless to say, whatever the *fungus* needs in order to psychologically stimulate manpower toward the generation of technological growth is what it develops next. Flow equals Grow.

BP Life on Earth started with a natural economy of matter/pure-energy, with the energy coming from the Sun being a constant driving input for the expansion of the number of systems engaging in this economy- a mechanism that allows the growth of consciousness without exposing the true origin of all life itself. As evolution began to produce more complex life, there came a point where life on Earth began to engage with that economy in a purposeful and appreciable way. This occurs shortly after the appearance of the BCT. The BCT goes on to significantly alter the way in which it interacts with the economy, as now, the organism has significant conscious control over the ways in which it exerts its influence over its environment. With continued time, technological advancement explodes alongside the fungal growth of the economy. New trade partners and trade routes, marketplaces, treaties, agreements, and other forms of societal integration accelerate the pace at which information is exchanged, resulting in an economic and technological boom. When the temperature funnel polarizes the global consciousness on Earth due to the awareness of impending demise, the cognitive motivation to reach TES no longer will come from money. This is the case because money can align the goals of certain groups or individuals in a manner that is contrary to the development of technologies that are required for TES. In fact, money as a motivator may become a distraction, slowing down the biological and technological progress that is required in order to achieve TES. Thus, what is required is that the entire collective consciousness of humanity has one focus, which is the development of technologies that will promote TES. This is a permeated collective consciousness (PCC) and is a requirement for all fungal growths to achieve this state in order to avert their impending demise. This is the difference between conscious-growths operating in an ACS versus a MCS.

BBP In all reaches of the cosmos, this mechanism of forcing fungal growths to reach a MCS by striving ever further toward TES in an attempt to both escape the increasingly hostile environment generated by their host star and to fill their VOP is universal. The specific mechanism by which all of this occurs is the formation of a bidirectional interaction between the psychology of the fungal growth and the matter/pure-energy economy in its biosphere. As the psychology of the fungal growth matures, an intermediary economy is formed to increase the efficiency of this interaction, and results in technological growth. This economy is one of a denomination-based exchange system. It is a necessary component of the advancement toward TES on any biosphere that should host a fungal growth.

Eventually, it is highly likely that these fungal growths that have reached a MCS and exist as a TCM will encounter one another. This will have a psychological impact on any being, for even if they had theorized other life might exist in the universe, actually finding it will have many reverberating consequences on their psyche. This will most likely be a positive influence, as intelligence is negatively correlated with violence due to the obvious logical outcomes of loss of life, resources, and technology exchange due to conflict. These entities will then go on to form economies that transcend the limitations of resource transformation and exchange on a single biosphere, opening up entirely new avenues for technological growth and information exchange. All of this occurs as the universe pushes these conscious entities toward MCG and prepares them for the UGCH.

EOS

- Pillage reduction
- Civility increase
- Bartering as a trust
- Banking as a regulator of civility
- Evolved technology

Intervention and its Effect on Psychological Evolution

TL;DR: The *fungus* is self-motivating in that it is always developing the stressors it needs in order to make good timing for the evacuation day. The *fungus* also relies on its external and universal environmental stressors that force the evolution and evacuation off its biosphere in a timely manner. It knows that it is critical in its development of the technology needed to become successful as a matured consciousness through wars, pandemics, mass extinctions, external forces, including information and/or from external sources. These are the incorporated fundamental limitations that make up the psychology that enforces actions carried out by the fungus' spore. That will ensure its success to MCG, not only biospherically but galactically as well.

The psychology of the fungus on a biosphere is shaped by many different interactions and events. Some are mundane in comparison to others, such as the color of skin, feathers, or scales that develop in a particular animal in different regions, as well as how that might affect mating and other behaviors. Some are much more impactful and are required for the biosphere to develop a fungus up to a spore stage that will be capable of evolving to meet TES. These interactions are called interventions, of which there are the following types.

Six Different Types of Intervention

1. Pandemic

ITM Throughout the course of evolution on Earth, various pathogens evolved to infect members of other species in massive numbers, causing mass casualties among those species. Such large-scale infectious conditions, or pandemics, pose existential challenges to the host species and stimulate adaptations among them that are vital to their continued existence. In the case of this occurring when humans have existed on Earth, many adaptations have taken place in order to ensure that the disease does not wipe out significant portions of the population, or potentially all of the population. The response to these threats (e.g. measles, The Black Death, and various venereal diseases) have forced changes in sanitation, medical developments in the form of treatments and vaccines, as well as further study of pathogen life cycles and adaptation responses.

BP It is extremely important that humans be threatened by biological agents or pathogens in this manner so as to advance the sciences of medicine, sanitation, and various imaging techniques.
These biological agents or pathogens then spread throughout the Earth and cause us to be brought face to face with the second fundamental limitation of the universe: inevitable death. Furthermore, our biological sciences have begun to rapidly improve in their understanding of the interconnectedness of all of the constituent organisms on Earth. As we study the various mutations of these biologically threatening organisms

and how they both improve and become more or less infectious with time, new information is gained that brings us one step closer to reaching ESS. If humanity is to survive the end of the Earth and achieve TES, it will absolutely require advanced medical understanding that came from biological threats to ensure its survival in outer space. This is the necessary biological development of a TCM.

BBP In order for the expanding universe to achieve its intended goal of MCG, this process must repeat on each and every biosphere. Each conscious-growth on each biosphere must evolve under the pressures and threats from biological agents in the unique composition that exists within each of those biospheres. The end result of this process occurring throughout the cosmos is the universe's intended end product of MCG, as these conscious-growths can go on to share their unique understandings and discoveries that sprang forth from their differing conditions as they evolved.

2. Threat of Impact from Outer Space

ITM Humans have been looking out into the stars for longer than exists in recorded history. However, it was only in the more recent parts of evolution that powerful telescopes were invented in order to give humans a truer view of what is going on in the cosmos. With these new observational powers came new concerns and questions. Prior to the invention of the telescope, when humans visited impact craters on the Earth, there was only wonder and confusion as to what could have caused these large scars on our planet. However, the telescope invokes a whole new realm of understanding that brings about the questions that change the psychology of all life on Earth. Did these marks on the land come from something falling from the stars? When humans began asking these questions, profound shifts in our sense of security on this little blue rock took place. Suddenly, the threat of being completely annihilated at any time became much more realistic and humanity was knocked down a peg or two in terms of its sense of superiority in the cosmos. This psychological revelation promotes a slew of behavioral changes that have pushed humanity toward reaching for the stars.

IPT In order to defend ourselves from comets and asteroids that could cause mass extinction, humanity first needs to be able to reach into space and escape Earth's security. Initially, this vision of escape was to be carried out by ballistic rockets. However, as humans evolved, we have set up stations that orbit the Earth and have traveled from the Earth to the moon. Each of these events ultimately occurred due to the sheer fact that our psychology was impacted by the thought that we might be powerless to defend ourselves should something come and slam into the Earth. There were other geopolitical issues occurring during these times, such as the race against Soviet Union To be the first in space, or to the moon; but these were all ultimately founded by the insecurities of humanity that were instilled by knowledge of our truly helpless nature.

More recently, NASA has devised a potential method for protecting humanity against future impacts using very basic means of colliding with an asteroid when it is far enough away from the Earth in order to redirect its path.

"NASA's Double Asteroid Redirection Test (DART) investigation team shows the spacecraft's kinetic impact with its target asteroid, Dimorphos, successfully altered the asteroid's orbit. This marks humanity's first time purposely changing the motion of a celestial object and the first full-scale demonstration of asteroid deflection technology."[116]

The importance of this type of capability cannot be understated. For example, it was only very recently that scientists were able to determine and confirm that the most recent mass extinction event that wiped out the dinosaurs was related to an impact event. While some life did survive this extremely devastating event, it is very likely that if this were to occur in modern times, it would wipe out most, if not all, humans on Earth.

BP The awareness of the threat of a potential doomsday impact on any biosphere, as well as the resulting inevitable increase and then subsequent decrease in global temperatures, assures that a conscious-growth, motivated by the psychology of fear, will develop myriad technologies. Some of these technologies may have the effect of bringing Earth's *fungus* closer to TES, whether knowingly or unknowingly. This particular intervention may also be the catalyst for a change in the *fungus*, where it begins the prep for the disbursement of spores by first burrowing underground. Then, while underground (The Subterranean Revolution, akin to the film *THX 1138*), the evolution of the *fungus* continues and the innovations for the final requirements for

TES to become a TCM are achieved.

BBP This eventuality is a terrestrial accelerator, which causes increased awareness in the psychology present on that biosphere. In essence, this psychology is forced to face the fact that it must meet TES in the limited time frame allotted for its respective biosphere or else perish. Given the fact that impacts are hard to predict and somewhat random in nature, the psychological effects of a constant threat of impact act to enforce a hurried or accelerated path toward TES. These threats could come in 10, 100, or even 1,000 years and the fact that it is rarely known far in advance is a constant psychological pressure pushing all conscious-growths toward MCG.

3. Threat of Self-Annihilation Anthropocene Epoch)

IPT There are a variety of proposed ways in which scientists believe humanity could end in self-annihilation. A list of these events and their associated probabilities is shown below (taken from a study conducted by the Future of Humanity Institute in 2008).

Risk	Estimated probability for human extinction before 2100
Overall probability	19%
Molecular nanotechnology weapons	5%
Superintelligent AI	5%
All wars (including civil wars)	4%
Engineered pandemic	2%
Nuclear war	1%
Nanotechnology accident	0.5%
Natural pandemic	0.05%
Nuclear terrorism	0.03%

Table source: Future of Humanity Institute, 2008.[23]

Figure x 11 (Extinction Probability)

This chart shows the estimated probability of human extinction to date.

Given that humanity has risen to a point where it is capable of numerous technological feats that can bring about the end of human civilization, new psychological considerations begin to arise. This is part of the development of awareness and conscious-growth within the *fungus*. Given that humanity has entered this stage, certain psychological impacts that come as a result must be analyzed in order to understand the evolution of the fungus to promote spore development.

One such analysis that can be performed is to examine how thoughts of death generally affect humans, even if the threat is only to an individual, and extrapolate outward toward larger threats.

"Terror management theory (TMT) proposes that thoughts of death trigger a concern about self-annihilation that motivates the defense of cultural worldviews."[117]

Terror management theory can be implemented in order to explain preservation behaviors in more than just worldviews but of self-esteem and interpersonal relationships as well. Ultimately, what the theory proposes is that people will behave in a way that allows them to find more meaning in their life in the face of death. This can have both positive and negative effects, depending on the individual and the type of threat.

"When death thoughts are readily accessible, and we feel threatened, this may cause us to become more close-minded and exhibit bias and stereotypes against others who don't hold the same cultural worldviews."[118]

These types of challenges are the kind that humanity must be forced to face and adapt to in order to be prepared for the ramifications of the temperature funnel. Simply telling all humans today that we have definitive proof that humanity is ending would cause immediate panic and anxiety. However, allowing such a terror to come in the form of potential annihilation events gives humanity time to come to terms with what it means to have evolved to this point. In other words, we are more able to understand our influences, capabilities, and responsibilities as we try to survive the very conditions we sometimes create. This is a gentler way in which the fungus eases humanity into achieving TES, minimizing the instantiation of biases and stereotypes, which often occur under societally imposed threats.

Given that terror management theory suggests this type of thinking promotes defensive behaviors toward cultural worldviews or the self-esteem of an individual, existential threat to all of humanity surely would be the largest promoter of such behaviors. Therefore, by mankind achieving the ability to destroy itself, it also brings about a set of psychological conditions that can promote the further preservation of life.

In part, this can be attributed to the terror of being unheard by others and/or invisible, both here on Earth and even in the grand cosmic scheme. In essence, at the greatest scale, it might be the case that humanity would be worried that should we annihilate ourselves, no one would have ever borne witness to our existence. These types of thoughts can elicit behaviors which promote protective effects for all of humanity. Wishing to innovate for "greener" sources of energy is one demonstration that is relevant in the modern world. Another example is wishing to preserve the life and material we currently have in order to restart life anew should some catastrophe occur on Earth.

Today there is a seed bank in the Arctic Circle that holds almost every seed on Earth in case of a global disaster.

"The vault is on the Norwegian island of Spitsbergen and contains almost a million packets of seeds, each a variety of an important food crop. When it was opened in 2008, the deep permafrost through which the vault was sunk was expected to provide 'failsafe' protection against the challenge of natural or man-made disasters."[119]

However, other behaviors are also exhibited that don't necessarily seem directly tied to a protective measure or solution. As evidence of this, humanity may also collect information for others to view and learn about us long after we have all gone extinct.

"If — or when — human civilization collapses, what will remain? A massive, indestructible box that'll record scientific data to give future civilizations insight on how exactly humanity fell."[120]

What can be seen is that the threats of self-annihilation evoke behaviors that promote some form of preservation. These behaviors are also occurring on a larger scale than just the individual level, as the many projects above demonstrate. Humanity, at large, is very aware of the possibility for itself to be completely wiped out. This is not something that was ever considered at a mass scale in earlier civilizations. This demonstrates the uniquely changing psychology of the *fungus* as it relates to its technological advancement and the threats that this advancement imposes. Of course, while these sound grim, they are absolutely necessary to prepare humanity for TES.

BP During this period of time, which is called the Anthropocene (anthropo- being related to mankind), man dominates the Earth. This dominance is a testament to the achievements and evolution of the BCT. However, it does come with a price. Having so much power that one can change the course of natural events on Earth means that this power comes with great responsibility. Humanity is learning lessons about how best to interact with its environment in order to survive long enough to achieve TES and exist as a TCM in ESS. The power we wield today exists as both a benefit and a threat to our very existence. This necessary stressor forces significant changes in the psychology of humanity and promotes further awareness. This can be seen in the behaviors of man in recent years.

If humans today have gone to the extent of preserving our biosphere through seed banks and recording information about human existence after it has been destroyed, then there must be either a survivor to reinoculate the planet, or an observer to bear witness to the recorded information for these endeavors to have any real meaning.

In the case of the seed bank, one can imagine a Johnny Appleseed (an American pioneer who introduced apple trees to large parts of Pennsylvania and the Midwest) on a massive scale. In the past few decades, man has created a station in space (the International Space Station--ISS) where experiments are run, ranging from watching how organisms grow in zero gravity to physical/chemical experiments. It would be ironic that the astronauts inside the space station can be viewed as the predecessors of the TCMs of the future if NASA had merely sent them to the ISS solely to conduct these experiments. Perhaps there will always be a human in space, forever inside, as the successor to the ISS.

In the more grim case, where humanity is entirely wiped out, the recorded information merely serves as an act in order to have our species be remembered. However, it may also serve to inform other species that may come along and detail to them a story of our *fungus* on this biosphere and hopefully provide some level of deterrence for any similar mistakes they might make. This, of course, is only the case if such mistakes are possible for a species advanced enough to have traveled from outside our solar system to uncover our recorded information.

BBP Each unique fungal growth throughout the universe will eventually come to a point in which its technological growth reaches a level where the use of that technology can cause biosphere-wide harm and threaten that fungal growth. Reaching this point is necessary for that fungal growth to become both aware of its biosphere's importance and begin facing the challenges associated with such influence. These are the beginnings of the development toward TES and ESS. Without facing these challenges, a fungal growth will be unprepared, through both a lack of experience and a lack of knowledge, to handle the complexities of cosmic existence and the requirements it imposes on self-sustainability. Thus, every fungal growth must reach a level where it threatens itself with annihilation and surmounts all obstacles associated.

4. Human-Influenced Climate Change

TL;DR: If it is that we as humans eventually need to leave Earth, then a little push in the right direction could be a necessity in achieving TES in the time allotted to do so. It is also fungal in nature in the timeline of evolution that man-induced climate change be a significant part in its evacuation success.

ITM Throughout the history and evolution of life on Earth, many different environmental changes caused by organisms have influenced the climate. These are mainly attributed to things like algae, bacteria, plants, and fungi. However, as mentioned prior, these changes to the climate were only due to the natural biological function of these organisms. While the climate on Earth has varied over geological timescales due to these organisms' influence and has looked vastly different than it does now, it is very much the case that the rate of change of the climate is accelerating due to human activity, and this activity is not part of the natural biological function of humans themselves.

"...ancient farming practices led to a rise in the atmospheric emission of the heat-trapping gases carbon dioxide and methane -- a rise that has continued since, unlike

the trend at any other time in Earth's geologic history. It also shows that without this human influence, by the start of the Industrial Revolution, the planet would have likely been headed for another ice age."[121]

These early farmers were completely unaware of their ability to influence the climate and therefore no psychological impact from agriculture could occur during this time. The fact that their activity did impact us, and we are more aware of that now, does have an overall effect on the psychology of humanity today. In fact, we have learned of this additional influence only after we have become aware of our impacts on the climate due to modern activities.

IPT At the start of the Industrial Revolution, the output of carbon dioxide began to skyrocket. The increase in the levels of carbon dioxide in the environment has caused changes in the atmosphere that are making the rate at which global temperature systems change sit outside the bounds of anything that has been seen in Earth's natural course of history. Storms are getting more frequent and more deadly, while the climate in many regions is also shifting so that areas that didn't have heat waves now do, which has been particularly notable in 2020-2022. The United States has seen some of its deadliest storms and fires as well as record numbers of these storms to occur in a given year. Europe has had roads melt due to record heat waves and floods have occurred in many cities that have rarely, if ever, had them. All in all, humanity is forcing itself to address these crises or perish, which only serves to act as preparation for TES. Despite the grim predictions for 2021, Nostradamus' prophecies continue until the year 3797. So if we make it through potential asteroid crashes, famines, the plague, and a zombie apocalypse, humanity might survive an extra 1,500 years or so. However, Nostradamus isn't the best or most credible source; therefore humanity should pay close attention to the climate as this is one of the greatest threats it is currently facing. In order to become the TCM and achieve TES, the current climate crisis needs to be addressed so that the threat to humanity only comes from the Sun and not ourselves. This extends the window of time for humanity to achieve TES quite significantly.

BP Human activity is one factor in how long the Earth may remain hospitable. However, it is inevitable that, whether it is from climate change that is greatly accelerated by human activity or the climate change that occurs due to the slow warming of the Sun, the Earth's climate will eventually be hostile to human life on the surface. There is also the confounding factor of the Sun's 11-year cycles in its activity that can obscure humanity's contributions to the climate. Regardless, humans will either be forced into The Subterranean Revolution to continue their development of technologies in order to ensure ESS by achieving TES or be forced off-world in the long run. Facing the changing climate is early prep for understanding the ramifications of the temp funnel and humanity will begin to understand the devastating impacts relying on our biosphere can have over long periods of time. This is the way in which changing climate begins to stimulate the requirements of ESS for the TCM.

BBP Since the universe's main purpose is to generate consciousnesses, and these consciousnesses should achieve MCG, climate change acts as an accelerant for the *fungus* on each and every biosphere in the cosmos. This particular motivating factor is baked into the design of star systems and the evolution of life itself. The abilities of increasingly intelligent conscious-growths to affect their biosphere become the initiator of the development of ESS throughout the cosmos. For, without negatively impacting their own ecosystem, they are unable to truly understand the value of it and the necessity to support themselves without it.

5. Overpopulation/Depopulation Cycle

TL;DR: Even the out-of-control human population is the course that the *fungus* will take when or if it is needed on this biosphere to succeed at its main goal TES.

IPT The Earth has experienced rapid growth in the human population over the last few hundred years. Currently, the population of the Earth is approximately 7.7 billion. The demand for food and resources is becoming unsustainable at these levels of population and is straining the strict economy of resources that exist on Earth. While innovations can make changes in the efficiency of certain chemical processes or production cycles, it is inevitable that these are merely stopgaps for a much larger issue. Humans are destroying forests, jungles, and various other lands that are essential in keeping the Earth within a specific set of bounds that will allow it to continue supporting life in the way we know today. Already, these various activities are stressing and strangling the systems the Earth has evolved in order to support life over billions of years. The level of conscious awareness and concern presently possessed by our collective consciousness is just beginning to polarize society and bring it nearer to a PCC state. The fact that this behavior is unsustainable is driving various movements, including affecting governmental policy on how much pollution companies can output, and is forcing governmental agencies to affirm that they will go carbon neutral by 2030-2050. This demonstrates that this acts as a psychological pressure and allows the *fungus* to further adapt to its own existence. The fungus will begin transitioning from "whorish" behavior to one of more sustainable coexistence with the ecosystem in which it finds itself. This is the case due to the second fundamental limitation of the universe: inevitable death. Should the fungus fail to find balance with the global resource economy, it will eventually perish due to the overtaxing of global resources and the destruction of its supporting ecosystems.

"Wildlife is dying out due to habitat destruction, overhunting, toxic pollution, invasion by alien species and climate change. But the ultimate cause of all of these factors is 'human overpopulation and continued population growth, and overconsumption, especially by the rich', say the scientists, who include Prof Paul Ehrlich, at Stanford University in the US…"[122]

Due to economic hardships, pandemics, and various harmful human activities, the recent trend in human reproduction has been the exact opposite of growing into overabundance. Humans had already reached an overpopulated state some time ago and the response to this overpopulation is a process that is decades in the making. The various activities of polluting, overcrowding, and over consuming the land, sea, and wildlife have put humanity in a decline for the first time in the history of modern man.

"What is more, over the past few decades, the quality of human sperm has declined massively, possibly leading to lower birth rates, for reasons nobody is really sure about. Pollution—a by-product of human degradation of the environment—is one possible factor. Another might be stress, which, I suggest, could be triggered by living in close proximity to other people for a long period."[123]

The first ramifications of this overpopulation and subsequent overconsumption and pollution of the Earth are just starting to be seen. The depopulation of the Earth is beginning due to the constraints on global supply and the quality of life on Earth itself.
"In most countries—including poorer ones—the birth rate is now well below the

death rate. In some countries, the population will soon be half the current value. People are now becoming worried about underpopulation."[124]

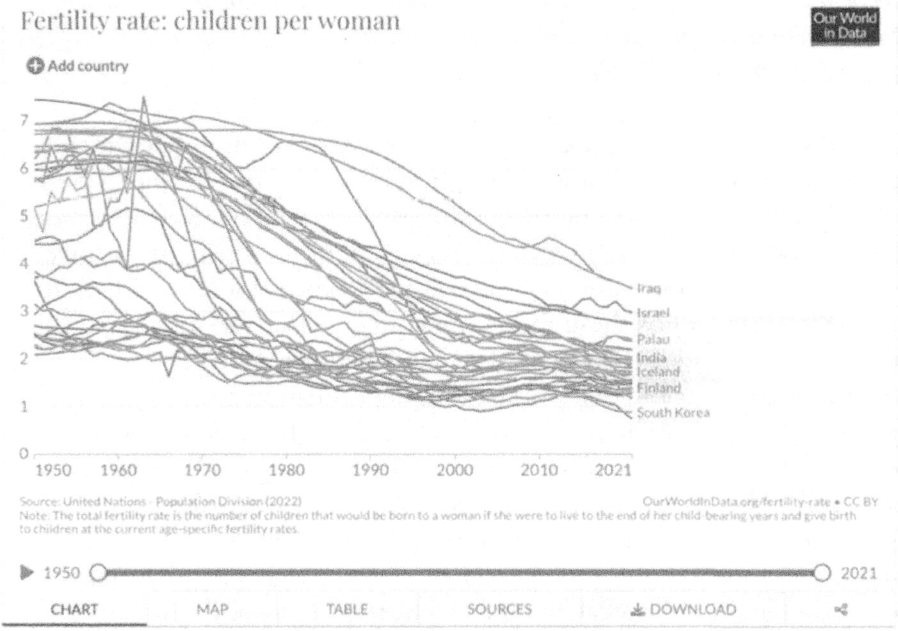

Figure W 41: (Fertility Rate)

This graph shows the birth rate decrease in the last seven decades and the possibilities of endangering the human race. The decrease is the result of the advent of birth control.

BP Earth's *fungus* is constantly adapting and evolving in order to achieve its end goal of MCG. However, prior to its achieving this state, it will have to adapt in a variety of ways. First, the *fungus* attempted to produce successful organisms. When the *fungus* has iterated a sufficient number of times, it will have achieved a particular bipedal conscious-growth which is extremely successful in terms of its ability to grow and thrive. However, this is insufficient experience for a TCM. The *fungus* is now experiencing its first signal that it has achieved a sufficiently successful conscious-growth that can expand globally and dominate the ecosystem. The ramifications of this are that the *fungus* will now have to adapt in reverse.

"The most insidious threat to humankind is something called "extinction debt." There comes a time in the progress of any species, even ones that seem to be thriving, when extinction will be inevitable, no matter what they might do to avert it. The

cause of extinction is usually a delayed reaction to habitat loss. The species most at risk are those that dominate particular habitat patches at the expense of others, who tend to migrate elsewhere, and are therefore spread more thinly. Humans occupy more or less the whole planet, and with our sequestration of a large wedge of the productivity of this planetwide habitat patch, we are dominant within it. H. sapiens might therefore already be a dead species walking."[125]

Earth is now seeing the *fungus* reach the stage of trimming down and paring back to become more efficient in order to price into its genetic code what is required to be successful for ESS as a TCM. This presents itself as being more sustainable, going green, or increasing the efficiency of the production of various goods, as well as novel ways in which to produce them. All of these things are concerns that are much more in the public eye today than they were half a century ago. This demonstrates that the overpopulation/depopulation cycle has a profound effect on the psychology of the *fungus* because it instills concern for the future of the species. This naturally lends itself toward a period of lesser growth and decreasing numbers in order to sustain humanity for longer periods of time. As such, overpopulation is a key milestone for the fungus to achieve in order to signal its next phase of evolution and a shift away from its "whorish" behavior. Humanity is just beginning its depopulation phase and will continue to do so for quite some time.

BBP Without these fluctuations due to overpopulation/depopulation in a fungus, it may be possible that a biosphere in the universe does not produce a conscious-growth that achieves its required TES within the time allotted for its evacuation date. This would likely lessen the number of fungal growths that have achieved MCG that are available for the UGCH. However, it will most assuredly decrease the quality of these fungal growths in terms of their level of awareness. In order to produce the highest quality fungal growths, the universe enforces a change in the behavior of the fungus that grows on any given biosphere. This is achieved through an effect on its psychology that comes as a result of adapting to overpopulation and subsequent depopulation due to resource strain.

Extraterrestrial Contact

ITM While humans have made reports of contact with aliens or beings from the stars for hundreds or even thousands of years, some in religious texts and others in military reports in more recent times, the takeaway from this is that these encounters are not so blatant as to allow all to agree upon the existence of extraterrestrial life. Any sufficiently advanced species that would have the technology required to visit the Earth (and acting as non-terrestrial ecologists) would understand the idea that interfering with a civilization before it has reached sufficient conscious-growth and psychological development would be reality-shattering and extremely detrimental to its progress toward TES (vandalism of consciousness), similar to how ecologists on Earth would be wary of interfering with the development of any subjects under their study. This is likely the reason for the sporadic and evasive encounters that have been documented throughout human history. There is also the possibility that the aliens will only step in and interfere with our natural development in the case of a severe disaster/catastrophe or if we are in danger of failing to meet TES. It is possible that discovering aliens of our own accord may be the most humane process for coming to terms with the realization that humans are not alone within the cosmos.

IPT A trustworthy piece of technology needs all the history behind it for that consciousness to understand and rely on it, especially for ESS. Since humans still have a large number of individuals whose entire identities and reasons for existing would be shattered by the forced reality that extraterrestrials exist, these galactic entities are likely labeling the Earth as a zone of "minimal interference." This idea is mirrored in the show Star Trek by Gene Roddenberry, where the "Prime Directive" dictates that non-interference is the most sacred law one must abide by when traveling through the cosmos where other primitive life may be taking root. This is exactly the same approach taken by researchers that encroach upon ape societies. Over many months or years, the researchers slowly introduce themselves into the habitats of these apes in order to increase their chances of successfully integrating into their societies for observation. If these researchers were to immediately enter into the surroundings of these apes, violence and chaos would ensue and observation would not be carried out.

Thus, interjections into our society can also be harmful if they come from civilizations that are far more advanced than ours. Roswell's UFO crash is one of the most notorious UFO sightings and supposed alien contact events that has occurred in more modern times. The government made it clear that they tried to cover up the events by switching their story in a very dramatic way and denying prior accounts of those involved, such as Mack Brazel's original account of the strange properties displayed by the metal wreckage that he had found on his farm or the statements initially made by lieutenant colonel Jesse Marcel.

Roswell Daily Record

RAAF Captures Flying Saucer On Ranch in Roswell Region

Claims Army Is Stacking Courts Martial

House Passes Tax Slash by Large Margin

Security Council Paves Way to Talks On Arms Reductions

No Details of Flying Disk Are Revealed

Ex-King Carol Weds Mme. Lupescu

Some of Soviet Satellites May Attend Paris Meeting

Roswellians Have Differing Opinions On Flying Saucers

American League Wins All-Star Game

Miners and Operators Sign Highest Wage Pact in History

Image of the Roswell Daily Record

July 8, 1947, issue of the Roswell Daily Record, announcing the "capture" of a "flying saucer." The news spread all around the globe.

The fact that the government retracted its statement so quickly and in a drastic fashion suggests that a profound psychological effect was either taking effect or predicted to occur in the general public. Either the government itself was scared to the point of denial, or they were worried about mass pandemonium occurring in the general public. In either instance, it is clear that the psychological effect of this event has demonstrated that it is one of the necessary interventions that must ultimately be addressed. It just so happens that it was deemed, one way or another, that humanity was not yet ready for this type of knowledge.

Glenn Dennis, the only remaining witness of the events that took place at the crash site in Roswell at the time was a mortician who reported that he had been contacted by the Air Force about providing small hermetically sealable caskets. Dennis mentions the request was canceled only a few hours later. Dennis also happened to be associated with a nurse that needed a ride to and from the crash site. Dennis mentions that she was scared and had witnessed an alien autopsy, as well as the fact that she was pacing around and vomiting as she frantically tried to inform him about the events. The nurse detailed what she had seen and frantically scribbled it on a napkin at a diner where Dennis had taken her after picking her up. The following images are copies that Dennis took of the drawings she made.

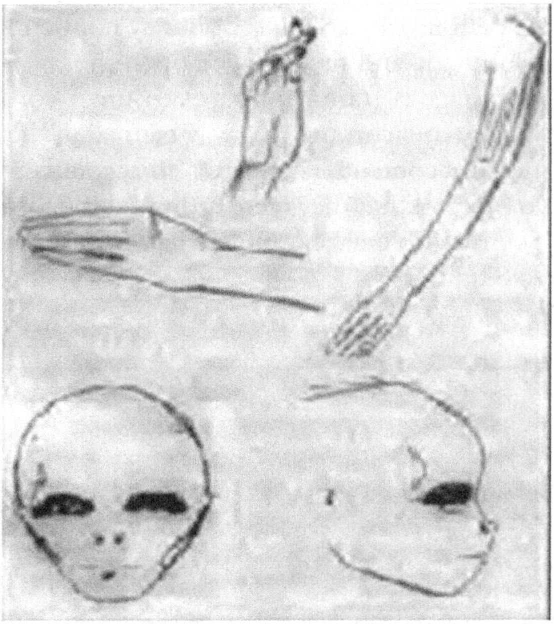

Figure U09: (Napkin From Nurse)

This image is a copy of the napkin the nurse drew for Dennis at a restaurant the night he picked her up from the military base in Roswell.

Dennis lost contact with this army nurse after she was shipped off to Germany and they exchanged one letter after she was overseas. After this point, all further attempts at communication resulted in a "return to sender." There is an interview with Dennis available online on YouTube at https://www.youtube.com/watch?v=_DA-g94Ro1I.

Another interesting piece of evidence is "Skinny Bob", a gray space alien from the Roswell incident that has been captured on film. This alien happens to have an appearance very similar to both Dennis' and the waitress' descriptions. The video first appeared on the internet in early May of 2011. The video became very popular in the UFO and alien research community, although it was speculated that the footage was a promotion for an upcoming film or TV series at the time. The video has also been presented as alleged leaked footage from the KGB, although this remains unconfirmed.

Skinny Bob

Skinny Bob was supposedly the only living survivor of the Roswell Flying Saucer incident in 1947. Skinny Bob may not be real; however, the film looked authentic and first showed up in the late 70s, not 2011. The Skinny Bob footage is 16mm and its exposure is consistent with the Black and White Kodachrome that was banned in the late 50s because of its mercury content. There are also accounts of those interviewed who claimed that they were around during the time in Roswell when Skinny Bob crashed. Notice how similar the napkin drawing looks to Skinny Bob.

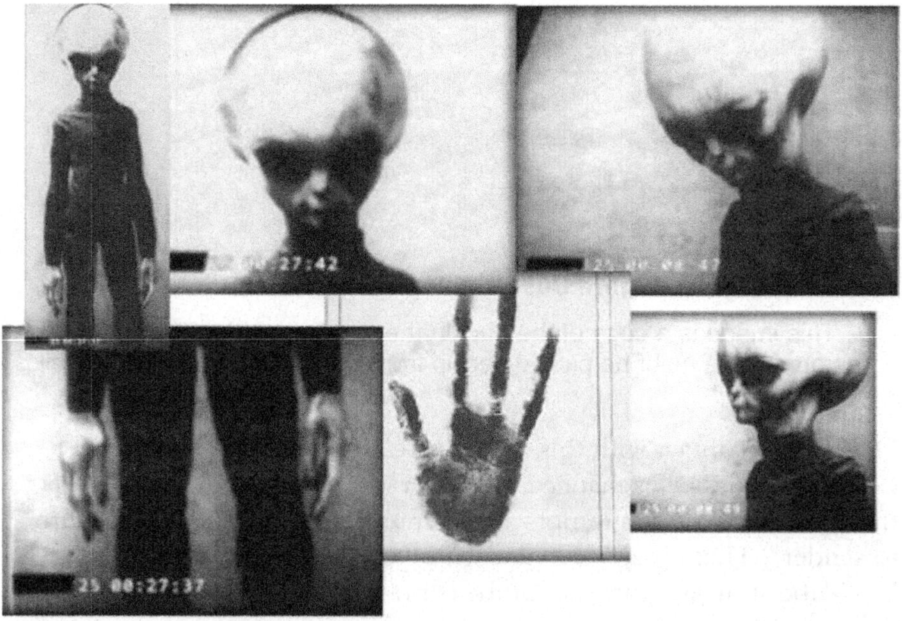

Figure SK1

Photos of the presumed and only surviving alien of the Roswell UFO crash in 1947, AKA Skinny Bob.

Another interesting admission came from the Vatican in 2008. The astronomer for the Vatican at the time, Jose Gabriel Funes, released a series of statements about the Big Bang as well as the existence of extraterrestrial life.

"Just as there is a multiplicity of creatures on earth, there can be other beings, even intelligent, created by God. This is not in contrast with our faith because we can't put limits on God's creative freedom."[126]

This admission seems almost random and unprompted. Generally, the Vatican does not concern itself with atheist views, particularly ones that relate to scientific principles of how the universe began or the presence of other life in the galaxy. This demonstrates a definite shift in the psychological perspective from the view of the Vatican about 500 years ago when Giordano Bruno proclaimed his beliefs about the universe being infinite and far greater than what the astronomers of the time would allow due to theological interpretations of the Bible.

"Italian philosopher, astronomer, mathematician, and occultist whose theories anticipated modern science. The most notable of these were his theories of the infinite universe and the multiplicity of worlds, in which he rejected the traditional geocentric (Earth-centered) astronomy and intuitively went beyond the Copernican heliocentric (Sun-centered) theory, which still maintained a finite universe with a sphere of fixed stars. Bruno is, perhaps, chiefly remembered for the tragic death he suffered at the stake because of the tenacity with which he maintained his unorthodox ideas at a time when both the Roman Catholic and Reformed churches were reaffirming rigid Aristotelian and Scholastic principles in their struggle for the evangelization of Europe."[127]

In 2020, The Black Vault (a website that now hosts over three million declassified government documents) made Freedom of Information Act requests to the Navy for declassification of UFO/UAP materials after the Navy released the infamous video of UFOs/UAPs captured by navy pilots on camera.

"More than two years later, the government responded with a letter that both confirmed that more UAP videos exist and denied the request to turn them over due to national security concerns."[128]

This demonstrates a shift in the psychology of the *fungus* as even naval/military complexes have admitted the existence and authenticity of these reports from their personnel. Given that there was only a denial to the release of the request for more videos and information, and not of the actual existence of such information, it can be concluded that UFOs/UAPs are real and likely much more known about by the naval/military complex than was previously alluded to at any other time in history. What can be seen is that people are being allowed to be made aware of this information very slowly over many decades. This is done in an attempt to prevent mass hysteria and preserve the psychological integrity of society. These admissions are the first step in moving humanity toward higher levels of conscious awareness and preparing us for the eventuality of life among the stars.

BP The influence of a definitive encounter with an alien civilization on Earth at some point before its end would have a profound impact on society. However, the limitations of communication between that intelligence and our own could come in a variety of forms. First, we may not understand any form of communication they attempt to employ. This has the effect of humanity trying to develop other forms of language that can be universally understood so that communication between ourselves and an extraterrestrial entity might be more successful.

A good example of this is shown in the film *Close Encounters of the Third Kind*, where the humans and aliens used a sequence of tones to communicate with each other. Another example is the film Arrival in which the attempt to learn the language the aliens used to communicate changed the consciousness of those who learned it such that they no longer perceived time linearly. Clearly, this particular consideration is beginning to pervade the *fungus* and ways to deal with it are being imagined.

Second, humanity may not understand the meaning even if they understand the words that are being spoken, either by some agreed-upon universal language or another form of communication. In this case, the results could be disastrous. The entities may share information and technology that ultimately lead to Earth's demise through its misuse. This would require that humanity reach a certain level of comprehension and maturity that comes only from experience and discovery. It is likely that these entities are capable of determining whether humanity has reached such a milestone by observing our culture and media. One such example would be that we would have admitted openly through various media outlets and announcements that we had discovered life elsewhere in the cosmos or that the reportings made here on Earth are indeed legitimate. The chaos and unrest that might ensue would take some time to resolve, and after this resolution has occurred, this might be the first time when these higher forms of intelligence step in.

BBP The distances between any two biospheres in the universe have remained quite vast for extensive periods of time. This demonstrates a mechanism of design that does not allow for constant interference from foreign civilizations, as well as contaminants that may come from a completely foreign evolutionary timeline. However, other intelligent species capable of traveling from one star system to another will likely be cognizant of the design of the universe and its inevitable production of conscious-growths. To that end, each civilization with these capabilities would likely wish to maximize the potential of developing life while still abiding by non-interference or minimal interference principles. As such, the act of interfering with the development of any biosphere is a potential tool for advanced civilizations to utilize only in the most extreme of circumstances. The fact that humans have video evidence of a potential extraterrestrial (Skinny Bob) allows one to look at the ideal GCT configuration. Skinny Bob's appearance is such that his small body utilizes less energy and most of the energy consumption is for an enlarged brain. Small appendages suggest that physicality is far less important, which is in line with what one would expect for an eternally self-sufficient society that exists solely in outer space. Since the universe is producing conscious-growths and housing their growth for sufficient time so that their consciousness is maximized, careful/sparing use of alien encounters as cosmological ecologists are a natural extension of the maximization process for conscious-growth, and ultimately, the UGCH.

Psychological Effects of Cognition on HTO that Promote Conscious-Growth: Emotions (Love, and Hate)

TL;DR: Similar to the interplay of light and dark, love and hate exist in a dynamic balance, each needing the other for recognition. You need love and hate both for love to be gauged as love and its hierarchy of values to even be noticed. The dominance of one over the other at different times allows us to objectively observe and define these values sociologically. Throughout history, this fluctuation contributes to the increasing development of technology that aids in understanding and defining these fundamental human values making us more conscious.

Mammals and reptiles that possess a brain exhibit differing levels of emotion and emotional complexity. There is the potential that the capability to possess emotions goes all the way back to the MCCTs, where specialized organelles were first evolved; however, this is not known. It is also possible that various forms of memory and memory storage instigated the evolution of emotions. Regardless, we see that emotions are something that has continued to evolve from the earliest stages of mammalian and reptilian life up until the BCT.

Humans respond to a variety of different stimuli and situations via complex feedback systems that occur biochemically, neurologically, and psychologically. These systems have evolved along with the *fungus*, and in our current protospore stage, some of these feedback systems represent what we now know and identify as emotions.

"Emotions are psychological states brought on by neurophysiological changes, variously associated with thoughts, feelings, behavioral responses, and a degree of pleasure or displeasure. There is currently no scientific consensus on a definition. Emotions are often intertwined with mood, temperament, personality, disposition, or creativity."[129]

In the case of the interventions mentioned prior, the emotion of fear dominates the mindset of humans and elicits behaviors that help humanity avoid the aftermath of these events, ultimately allowing humanity to be more likely to reach TES. Fear is only one such emotion, which then demonstrates that the evolution of conscious-growth is maximized by at least one emotional state. Therefore, it is highly likely that emotions, in general, are necessary to reach MCG. Another thing to note about emotions is that they occur faster than the speed of rationalization. By having responses that are elicited by stimuli in the environment, rather than by excessive amounts of rationalization, a consciousness can respond much faster to the events of daily life. Without these more complex feedback systems between the biology and chemistry of an organism, MCG cannot be achieved. Similar to the concept that a being that is aware of its true origin will have no impetus to act or discover, a being without emotion will not be driven to respond and prepare for the harbingers that may lie ahead.

A. Love

The emotion of love is another facet of the emotional response that is generated by a sufficiently complex growing consciousness anywhere within the universe. What changes in each later stage of a *fungus* is the complexity of how the emotion of love is displayed. Love is an emotion that evolved throughout the development of life on Earth as a mechanism that confers greater chances of success to offspring through protection given by filial bonds.

"Reptiles show minimal parental investment in offspring and form non-selective relationships between individuals…By contrast, many mammals show intense parental investment in offspring and form lasting bonds with the offspring. Several mammalian species—including humans, wolves and prairie voles—also develop long-lasting, reciprocal and selective relationships between adults, with several features of what humans experience as 'love'. In turn, these reciprocal interactions trigger dynamic feedback mechanisms that foster growth and health."[130]

This likely occurred in these animals because animals generally have fewer offspring compared to earlier forms of life that lay large numbers of eggs and utilize these large numbers of offspring in order to ensure the success of at least a few to perpetuate the species. Thus, love is an emotion that promotes the survival of creatures that have fewer offspring in order to increase the chance of survival.

Some degree of love must exist later in the CC of the *fungus* in order for there to be a reason to persevere. In more advanced forms of consciousness, love can motivate learning and discovery. Through this type of love, innovation occurs and technology progresses. Therefore, love plays a pivotal role in ensuring the continued survival of intelligent organisms, as well as allowing the fungus to iterate toward a perfected spore. The other types of love that can come in more advanced forms of consciousness are the love of oneself, of life and existing, and that of others in a romantic sense, or for friendship. Each of these other forms of love that developed with greater evolution of the BCT is fueled by the VOP.

ITM Humans demonstrate some of the most complex displays of love, whereas dogs demonstrate a more primitive form of love. Individuals who separate from the collective and face shame and humiliation in order to pursue various aspirations (usually ones that deviate from the norm) display a level of love for the collective that is more than the average level of love that is displayed by the collective toward any one individual or group. This display of greater love is not possible for organisms that are lower in complexity than humans. This is mainly due to their inability to communicate with as much granularity, which affects their social structure as well. To understand this concept as it relates to evolution, it is important to view love from the vantage point of complexity. As mentioned before, the Tree of Fungal Mutation shows the varying levels of complexity in the evolution of the fungus leading up to humans. The trunk of this tree was a through-line for increasing levels of concern. In order for a consciousness to display love, it must possess concern. Greater levels (or increased complexity) of love go hand-in-hand with greater levels of concern, as a conscious entity must have a love of life and of living in order to be concerned about losing its life or the lives of others; there is greater value in life.

IPT We are all aware of filial love as it is deeply ingrained into our society by this point in evolution. Parents display the deepest love imaginable for their children and are willing to sacrifice themselves in order to ensure their survival. This serves the purpose, as described before, of ensuring that the race continues to survive and the next generation that has the potential to be smarter, faster, and stronger can survive.

In the case of an individual, love can be displayed in a number of ways. Primarily, the love of existence keeps each individual alive, even those that are not in the best of circumstances or aren't particularly skilled, or are suffering unfortunate circumstances. These individuals possess the self-love that is required for such consciousness to continue driving forward and continue existing. This is in contrast to animals that are less developed in their psychology and brain anatomy and are not as able to rank themselves or their skills in relation to the rest of their species or clique. Such thoughts of being inadequate would not fill their waking moments nor would they be presented with these facts as most members in animal packs must fight for their survival, thus proving their worth continually.

When primary needs are met, humans can display love toward one another in a romantic sense. This type of love is not necessary for the survival of the species. However, it does civilize the societal structure of humanity in some sense. Compared to reptiles and other organisms where selective breeding is not seen, this behavior additionally promotes a connected family unit that can help the survival of offspring.

Overall, there is some sense of communal love for each member of the community that has developed throughout evolution, where each member of the collective has some baseline level of care and concern for members of their society, such as country, state, or city. This also promotes a more cohesive structure in society and allows people to work toward a common goal together.

An example of such a concerned individual displaying love for the collective that wishes to propel technology forward due to his awareness of the temperature funnel would be Elon Musk. He continually attempts to propel humanity into a new era as a multi-planetary species by starting with the colonization of Mars in order to preserve our consciousness and escape inevitable death just a bit longer.

BP Since every BCT will be influenced by the VOP, they will attempt to fill the void via the mechanisms described previously. Since all forms of consciousness are part of the collective, the action of BCTs filling the VOP makes them a greater contributor to the collective, which ultimately increases their capability of spreading and receiving love. Love increases in value along with the increase in consciousness that the *fungus* produces. A BCT that does not engage with the greater collective will be incapable of receiving and sending emotions throughout the fungal network, which diminishes its capacity for love.

The periods in which humanity pushes forward at a rapid pace are generally brought on by conflict and war; however, innovations and discoveries usually occur in times of peace where inspiration has time to flourish and the love of innovating is allowed to take root. Since technology is spawned from invention and innovation, which has direct roots in necessity and love, the following process demonstrates how technology is created through emotion, specifically love, in order to reach TES.

BBP Once a conscious-growth has met TES, they will have reached the end of their adolescent conscious state (ACS) and continue on to achieve ESS as a TCM. It is at this stage that they are considered to be in a matured conscious state (MCS). At this stage in the conscious-growth's evolution, love expands outside the realm of being related to one's own species (altruism) and the species that existed on that growth's biosphere to all species in the universe, regardless of their present stage of evolution or, fungal growth. Conscious-growths that have reached this level of awareness will realize that both the rarity of finding a biosphere capable of hosting complex forms of life and the actual formation and evolution of that life demonstrate that all forms of life are precious. Such entities would then wish to foster the development of all life, respecting the sanctity of its development so as to not vandalize or taint that development. This is where the idea of intervention (appropriate levels of intervention) becomes important in the development of conscious-growths on any biosphere.

At this point, it will be entirely thanks to technology that one will be able to love. This is largely due to the fact that to exist in this stage, gender will likely be done away with and advanced gene editing technologies will generate what could be considered "offspring." Thus, the rules regarding relationships and connectedness between individuals will be much more evolved when compared to the relationships we are accustomed to today. Entities that exist under these conditions become galactic nurturers in the garden of the cosmos, ensuring that all life in the universal plantation of biospheres is allowed to thrive in order to reach MCG and contribute to the UGCH.

B. Fear/Hate

TL;DR: Fear triggers avoidance behavior, causing either individual survival or harm reduction. Hate can arise from fear, absence of love, or as a learned response opposing love. In the *fungus*, hate functions to destroy and diminish, stimulated by the fear of losing vital resources. Initially used for social structure awareness, hatred became a means to address issues. Biologically, love and hate share brain regions, but the patterns of activation are unique. Hate, less associated with deactivating rational centers, involves contemplation and rationale, though rationalizations may be flawed. Hatred's intensity can indicate genuine concerns or highlight informational deficiencies in societal structures, depending on the rational basis for its expression.

"Hate is primarily defined as a secondary emotion that arises from fear, including fears of rejection, abandonment, suffocation, and mortality. Clinicians may approach the experience of hate in terms of its interpersonal function, rather than a problem to be reduced or eliminated."[131]

Hate can also be a secondary and learned response in opposition to love. An emotional response of hate can form when either love is not present or is present in insufficient quantities. Under these conditions, as well as potentially others such as neglect, hatred can form. Hatred is often destructive and malicious and it begins to dominate the general quality of experience in that conscious entity after the feelings of love have festered or are not present.

ITM The function of hate in the *fungus* is to destroy and diminish. While this response can be generated by a lack of love or the presence of fear, it has the effect of stimulating the *fungus* to act. For example, hatred can arise from the fear of losing resources that are vital to survival, the second fundamental limitation of the universe: inevitable death. This would promote an attack on those that oppress an individual or group that attempts to take key resources from another individual or group. Hatred is much like a stressor in evolution, which forces an adaptation, except hatred arises from a stressor on the consciousness and not on the physicality of a conscious entity. This stressor alerts the individual or collective that something is wrong or suboptimal and that action needs to be taken. If only love existed, and a sense of complacency was to set in, the fungus would not act at all, as concerns, being related to fear, would not exist. This would have the effect of endangering the fungus' main goal of achieving TES.

Initially, hatred was used by the fungus in order to bring about awareness of an issue in the social structure of the group. For example, in many primates, behavior that represents hatred was exhibited by the alpha males to assert their dominance over other/beta males. This had the effect of letting the collective know which member was the strongest and fittest for reproduction. In essence, the act of hatred is a way that these animal communities are able to demonstrate the issue that other unfit males are attempting to mate, which would weaken the overall population. This increases the overall fitness of the group over time. This behavior carries over into more evolved primates including the BCTs.

"Hate is often considered to be an evil passion that should, in a better world, be tamed, controlled, and eradicated. Yet to the biologist, hate is a passion that is of equal interest to love. Like love, it is often seemingly irrational and can lead individuals to heroic and evil deeds. How can two opposite sentiments lead to the same behaviour?"[132]

Interestingly, the regions that control the emotion of love share a significant overlap with that of hate. These regions may fire together, but the pattern of that firing is unique between love and hate.

"A marked difference in the cortical pattern produced by these two sentiments of love and hate is that, whereas with love large parts of the cerebral cortex associated with judgment and reasoning become de-activated, with hate only a small zone, located in the frontal cortex, becomes de-activated."[133]

Since hate is an emotion less associated with the deactivation of rational centers, it is instructive to view the thoughts and concerns of those exhibiting hatred in a collective consciousness.

Hypothetical 1:
If someone is entirely filled with rage or anger, they will often devise strategies in a devious and rational manner in order to exact their revenge.

Hypothetical 2:
When someone is madly in love with someone else, all rationality flies out the window and will wish to commit acts of kindness and shower them with gifts regardless of any sense of reason.

Since the centers of the brain for rationalization remain active, hatred is an emotion that comes with some level of contemplation and rationale. However, it does not mean that the rationalizations need be true or reasonable, as rationalization can fail if the information held by the individuals or groups rationalizing is imperfect. Thus, in times where hatred swells due to rational and verifiable reasons (think about the hatred toward Hitler during the Holocaust), there is an indication that truly heinous acts are taking place and the hatred is aiming to alleviate the situation in such a way that is focused on the perpetrators of such terrible actions. However, in the cases where rationalization fails due to improper upbringing or inaccurate information, and hate begins to swell, there is evidence that information is not properly being conveyed to groups of people. This indicates either malicious intentions of various actors, the failure of the educational system, or some other form of suboptimal informational exchange in the *fungus*.

In either instance, hatred allows insight into the state of the *fungus* as hatred swells, if an adequate inquiry is performed surrounding the thought process of those engaging in hateful behaviors.

BP These exhibitions of fear, which often leads to hate, within the collective consciousness demonstrate the process by which awareness grows within the *fungus* and are the stepping stones that are required for the time when the collective consciousness on Earth is polarized by the temperature funnel. Given that these emotions are bringing attention to various concerns and issues that are brought about by individuals or groups, this particular function of emotion strikes a balance between stagnation and progress. In essence, being made aware of a problem through hatred removes the possibility of stagnation because the solution often requires change, which often also requires technological development and advancement. The use of technological innovation to solve problems becomes more and more common as the *fungus* evolves.

BBP Hate's use across the entirety of conscious-growths throughout the cosmos is essentially the same. This emotion acts as a spotlight to emphasize issues that exist within a dynamic interaction between individuals, groups, and even interspecies interactions. As fungal growths explore the cosmos, they may encounter other fungal growths. If one or both of these fungal growths have not achieved ESS, then hatred may still play a role in their interactions to highlight issues or concerns that one or both groups may possess. In this case, continued displays of hatred force the groups to analyze these behaviors and emotions, as the consequences for not doing so are dire. However, it is highly likely that any galactic civilization, that would have the chance to encounter another out in the cosmos, would have achieved ESS and exist in a MCS. As such, the second fundamental limitation of the universe is minimized and the usefulness of hatred would be significantly diminished or even entirely eliminated. If there is no issue in generating resources for oneself or others with whom alliances may be formed, there is far less chance that any concern could not be immediately alleviated through graceful cooperation. Thus, hatred is likely only useful to fungal growths up until a certain point of maturation.

Those existing in a MCS may still encounter those that have not reached a MCS, where acts of hatred from the lesser evolved target the more advanced conscious-growths. In this type of encounter, if the lesser evolved conscious-growth that exists in its ACS is either exhibiting large amounts of hatred toward its own kind or toward the more advanced conscious-growths, this indicates that the lesser evolved conscious-growths are not ready to interact and have mass awareness of the more evolved conscious-growths. Once again, hate has highlighted an issue.

How the Interplay and Dynamics of the Different Emotions in Society Promote Conscious-Growth

Love & Fear/Hate

TL;DR: Emotions are reactions; as an individual they are positive feelings from the possibility of future events or creations. Emotions are also negative due to fear and shame from that individual's sociological perspective. These cognitive self-identifiers are what cause us to act at all; without it we would remain still and silent. It is fear that gets us to conspire, protect, investigate, evaluate, record, and project.

The simplest analysis of an interplay of emotions in society would be the interplay between love and hate. Other emotional interactions that generate unique experiential states are more complex and often involve combinations of emotional states. In order to understand the many behaviors of a fungal growth, one can investigate the interaction of the fundamental emotions of love and fear/hate.

ITM As various ape and hominid species began to evolve, picking up on emotional cues became more important, which is evidenced by the fact that these ape and hominid species began to gain more facial control for expression.

"Humans and great apes are highly social species, and encounter conspecifics throughout their daily lives. During social interactions, they exchange information about their emotional states via expressions through different modalities including the face, body and voice. In this regard, their capacity to express emotions, intentionally or unintentionally, is crucial for them to successfully navigate their social worlds and to bond with group members…Compared to apes, humans have evolved communicative faces where all the expressive parts are emphasized."[134]

This demonstrates that emotionality and expressivity are both linked and part of an evolutionary process that the *fungus* continues to progress through in order to reach greater states of consciousness and awareness. It also demonstrates the earlier point that emotional processing is often utilized instead of rationalization or verbalization as it delivers information without requiring deliberation. Emotional processing occurs rapidly and reveals to others in the collective a variety of both personal and interpersonal information. Ape species are easily able to understand many different situational circumstances without vocalization merely by reading the body language and facial expressions of members in their group.

As emotions continue to evolve, the complexity of their interaction and use brings about new concerns, new problems to address, and also more efficient communication in general. As the *fungus* progresses toward a BCT, emotional centers and the musculature that supports the expression of these emotions will have developed significantly enough to produce a complex bevy of emotional states and communication modalities in order to ensure that problems can be effectively communicated and worked upon by the group. This has the effect of preparing the *fungus* for the creation of technologies, as these are the requisite faculties to possess in order to achieve such tasks.

By the time the BCT is gathering in large social groups and generating technologies in order to sustain and defend itself, the dynamics between love and hate will follow an undulatory pattern. The figure below demonstrates how this pattern evolves through different periods of time in which innovation occurs and technology is generated.

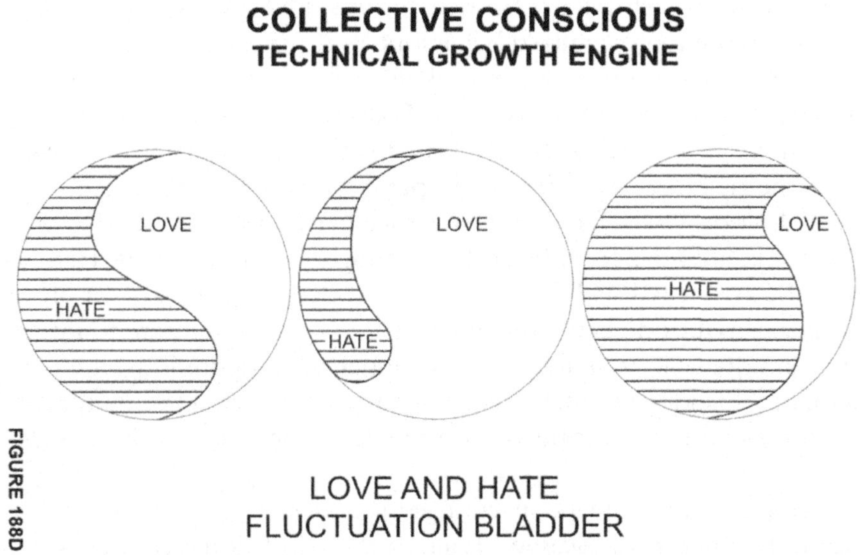

Figure 188D: Love & Hate Fluctuation Bladder

This diagram shows how love and hate work together as an engine for technological growth. By one temporarily being more dominant than the other, technological advancement occurs through two different mechanisms.

1) As hatred begins to swell and dominates the love displayed throughout

the *fungus*, those that exhibit love are tasked to resolve the threat which generally requires the generation of new technologies.

2) Conversely, if love begins to swell and dominates hatred in the *fungus*, then there is little-to-no sense of urgency or adversity to overcome.

In this state, a calm and collected approach toward discovery and innovation is fostered. When balanced, this system coincidentally looks like the symbol for yin and yang, which is also a balancing belief structure held in Chinese cosmology.

It is important to note displays of love or hate may be different from species to species. However, there are general trends in the behaviors that are conserved. Hatred is not associated with nurturing behaviors and love is not associated with violence or neglect at the most general of levels. Given the fact that these emotions can be expressed with some variation from species to species, some conventions must be used in order to describe certain behaviors that modern BCTs display as being either one of love or of fear/hate. Further analysis of the fluctuation bladder will be helpful in understanding its role in the advancement of technology. In such analysis, love and fear/hate will be broken down into generalized behaviors, as those are the only observables we can utilize when looking at the past, that are exhibited consistently when the majority of the collective is operating under that emotion. Love will mainly be associated with the concern for the loss of life, the lack of pride being the impetus for action (national, personal, or otherwise), a desire to reduce harm or protect, and existence in abundance without the wanting of vital resources. Hate will mainly be associated with the lack of any concern for the loss of life, a strong sense of pride being the impetus for action, the desire to inflict harm, and the lack of vital resources. Under these definitions, the fluctuation bladder can make sense of various historical events that were profound in their influence on the development of technologies.

A collection of individual conscious entities makes up the sum total state of the fluctuation bladder. Each of these individual entities is committing a variety of actions, each of which contributes to either the accumulation of love or hate. Various changes to the general milieu in which these entities exist can influence trends in the behavior of the individual entities. However, it is the overall pattern that determines the overall state which can be perceived at the macroscopic level.
IPT Individual conscious entities commit acts of hatred for a variety of

reasons. However, these acts are generally associated with beneficial outcomes for those entities that commit these actions while being at the expense of other conscious entities. These actions of hate fall into a number of categories:

Murder
- Abuse and/or to convince another to commit suicide
- Actively murder another to prevent the discovery of certain truths

Murder in retaliation
- Manipulation
- Gaslighting
- Triangulating
- Exploitation

Controlling Behaviors
- Threatening with abandonment or suicide if certain conditions aren't met
- Restricting interaction with other conscious entities
- Financial or material control

Other actions that are associated with love fall into other categories and generally promote the well-being of other conscious entities. These actions of love fall into the categories below:

Healing
- Curing cancer or other diseases
- Taking in and caring for the injured or abandoned
- Listening to the problems of others to help ease burdens and mental suffering

Supportive Behaviors
- Positive reinforcement in order to help others complete their goals
- Providing resources and advice to help others achieve their goals
- Providing a safe environment for others to grow or achieve success

Charitable Behaviors
- Giving money or possessions to help the less fortunate
- Spending time with the less fortunate to give them a sense of belonging

These groups of actions are usually responses toward individual conscious entities that are from opposing categories of love or hate. An entity that commits acts of hate will often cause some form of trauma that can be alleviated by an entity that commits an act of love that can alleviate the harm caused by the entities committing acts of hatred. While very general, this system is constantly at play. Here is one example of how love and hate work together to propel technology.

During WW2, Nazi Germany developed the V2 rocket in order to gain a military advantage over their adversaries of which they were the aggressors. Germany was not concerned with the loss of life of their adversary and thus they aggressed due to pride (a belief that Germans were a superior race) and the stresses imparted by a lack of resources (fuel and oil were severely lacking). Wernher von Braun worked on the V2 rocket from a place of love for innovation and was mainly apolitical. However, the German government was at war and their use of his technologies and abuse of his naivety allowed his work to be utilized for hateful purposes.

"Von Braun was motivated by his childhood obsession with spaceflight, a somewhat uncritical patriotism and a naive grasp of the ramifications of his actions in creating one of the War's deadliest weapons."[135]

The general populace was much more easily brainwashed by Hitler to believe in the ideologies of superiority, due to the fact that the country was in the middle of a depression. Those who were not brainwashed were intimidated by fear of being jailed or killed should they speak up against the regime. If one were to imagine the bladder which contains the amounts of love and hate that exist at the global level, hatred had been swelling in Nazi Germany since their loss in WWI. Hitler was framing this loss on a small group of German Jews, as well as sowing the seeds for his campaign of hatred to run on. This campaign was one in which the Aryans were the superior race of humans on the planet. By the time WWII had begun, hate had swelled in this bladder to a level that it had created a PCC. The development of technologies was accelerated by those existing in a state of hate in order to eradicate their foes.

The rest of the world responded with love–the love of life and liberty for people from all walks of life–in order to resolve the conflict and bring about a period of peace. In the interim, the development of technologies to thwart the Nazi regime was accelerated in response to the swell of hatred that had been taking place up until this point. During this war, the U.S. was fighting against not only Germany, but Japan after it had attacked Pearl Harbor a few years after the war with Germany broke out. The U.S. was very concerned with a prolonged war by this period of time, as the war had already been continuing in Germany for roughly 2-3 years when Japan attacked Pearl Harbor. Development of nuclear weapons accelerated as the U.S. planned to make a decisive strike to end the war. Here, the U.S. was operating under the concern for loss of life as well as the desire to reduce harm. The U.S. was not defending itself due to any pride it held as being superior to another country or people but merely a defense in general, as it was not the aggressor in the war. Thus, the U.S. was operating under a greater emotion of love and protection of its own people, and likely, the world. Ultimately, the U.S. dropped the first nuclear weapons on Japan in Hiroshima and Nagasaki. A few short days later, the war was declared over as the U.S. had already defeated Germany a few months prior and Japan was scared of continuing the war. This display of strength reduced the total number of casualties and economic losses to both sides.

"Because of the extent of the devastation and chaos—including the fact that much of the two cities' infrastructure was wiped out—exact death tolls from the bombing of Hiroshima and Nagasaki remain unknown. However, it's estimated roughly 70,000 to 135,000 people died in Hiroshima and 60,000 to 80,000 people died in Nagasaki, both from acute exposure to the blasts and from long-term side effects of radiation."[136]

In order to understand how this massively devastating event could possibly be a form of love in order to save lives, it is important to look at the loss of life that would have occurred if the war were to have dragged on. Given that the two bombs killed between 130,000 and 215,000 people, the war would have to have contributed to at least this many deaths in order for these bombs to be justified as an act of love in order to save lives.

"The March 1945 firebombing of Tokyo alone killed some 120,000 Japanese. A ground invasion would have resulted in nearly immeasurable casualties. As one scholar who studied the U.S. invasion plan, Operation Downfall, notes: "depending on the degree to which Japanese civilians resisted the invasion, estimates ran into the millions for Allied casualties and tens of millions for Japanese casualties."[137]

If one flyby of scatter bombs caused 120,000 deaths alone and there are estimates that continued ground invasions would have caused upwards of many millions of deaths, there is enough proof to suggest that the bomb did end up saving many more lives than it took. This only considers Japan and America in terms of loss of life. When we look at other countries that were participating in the war, this number grows larger still. Thus, the number of lives saved is likely at or around a minimum of 10 million people saved by dropping the bombs on Nagasaki and Hiroshima. This also excludes any lives saved from the realization that nuclear weapons are devastating and should not be used ever again. If the atomic bomb had not been dropped at this point in time, other nations with a larger arsenal of more powerful nuclear weapons would have chosen to use them at a later date and the destruction would have been catastrophic. The dropping of these bombs represents a point in time when the *fungus* existed in a love PCC, and technologies had rapidly been developed in order to end further demonstrations of hatred and the subsequent loss of life that is a result.

If everyone existed in a state of constant love, there would be no urgent need for technology to grow and ultimately meet the demands of TES. The same can be said for a state of constant hatred. It is the struggle between undulating levels of love and hate, as well as emotional states that are combinations of these two emotions, that changes the paradigm in which technologies advance between one of invention, the design of something completely novel when love dominates, and the accelerated improvements upon pre-existing technologies in the form of improvement that mainly occur when hate dominates.

BP It is the love and hate dynamic that is responsible for the empathic view of conditions of others during the terrestrial evolutionary process. Without this condition, technology would become stagnant and the *fungus* may not reach its TES goal.

BBP Throughout the universe the calmness of living in space, which may have its advantages, is similar to the need to leave water for land in order to live in more civil conditions.

Combinations of Emotional States

TL;DR Relationships serve as significant identifiers for conscious entities experiencing the Void of Psychosis (VOP), filling the void with a sense of purpose. The breakdown of a relationship can lead to feelings of despair, contributing to the Evidence of System (EOS) for the VOP. This emotional experience is a key motivator for actions that drive technological advancements.

ITM Organisms that display emotion are often more complex in the sense that they don't exhibit a pure form of one emotion or the other at a given time but an interplay between emotions that generate a unique quality of experience. Therefore, the truth about emotions is that they are gradients and there are shades of gray between each unique emotional state.

The amygdala is associated with fear and violence in reptiles and other primitive organisms. However another part of the brain has been developing in order to mitigate the signaling from the amygdala in order to control these emotions throughout the evolution of early hominids and humans. It is this region that affords the ability to overcome fear and violent tendencies in order to generate a contrasting state of empathy. Thus, empathy and narcissism can be thought of as two psychological states that are brought on by emotional influence that are on opposing ends of the spectrum.

The increased ability of more evolved hominids to override their primal fear/hate response demonstrates that when compared to earlier primates and the Neanderthal, larger social groups have selected traits that allow modern humans to be able to cope with individual differences and interpretations of experiences. This coping mechanism is forcing the growth of specific regions of the brain within the cerebral cortex and potentially other areas. As this region of the brain grows, empathy increases which also raises the level of general awareness. In general, this demonstrates that the fluctuation bladder not only influences technological growth but develops human consciousness and awareness.

IPT Humans display vastly more emotions than that of love and fear/hate. However, these are the most powerful driving forces in terms of technological advancement in the *fungus*. In the early psychological development of modern humans after their birth, the primitive brain is pre-wired; however, the rest of the wiring occurs through experience and learning. This pre-wiring relates to the strong contribution the amygdala has in generating emotional responses. Once again, this part of the brain is associated with violent behaviors and fear responses. In individuals with genetic predispositions toward suboptimal development in certain parts of the brain, the ability to develop more emotionally intelligent coping mechanisms, more resilient personal belief structures, and the ability to generate validation internally become impaired. In this case, empathy is not able to be displayed due to physical limitations. In other instances, a developing child will have poor role models and unfortunate experiences, which lead to the development of other patterns of firing in these structures associated with hatred, fear, and inadequacy, instead of the firing patterns of other regions that can regulate the emotions of fear, hatred, and inadequacy.

"Drawing on the greater impact of inescapability over unpredictability in animal models, we propose a mechanism by which ELA aggravates an exhaustive cycle of amygdala expansion and subsequent toxic-metabolic damage. We connect this neurobiological sequelae to psychosocial mal/adaptation after ELA, bridging to behavioral studies of attachment, emotion processing, and social functioning."[138]

In both of these cases, the child has a much higher likelihood of developing narcissistic tendencies and/or narcissistic personality disorder.

"Researchers used magnetic resonance imaging to scan the brains of 34 people, including 17 individuals who suffer from narcissistic personality disorder, and found that pathological narcissists have less gray matter in a part of the cerebral cortex called the left anterior insula...The left anterior insula region of the brain, which is thought to be involved with cognitive functioning and the regulation of emotion, has also been tied to the generation of compassion and empathy."[139]

Narcissism is the inability of one individual to feel empathy for others upon considering the other individual's circumstances. It also relates to strong senses of personal insecurity, which require consistent forms of external validation in order to quell these feelings. It is a narcissistic bubble that all animals must have in order to protect themselves from hostile environments and it is only when one can burst out of that bubble that it forms environmental concern.

The increase in feelings of insecurity can be seen in the way society has changed over the last three-quarters of a century in terms of narcissistic tendencies displayed by people and the effect it has had. In the 1970s, narcissism was less prevalent than it is today. However it was significantly higher than it was in the 1950s. Since society was not as aware and socially open during this time, cases of narcissistic abuse were not recognized as a common issue, causing many people to stay quiet in fear of shame, go insane or kill themselves. Because of this, levels of narcissism may have been underreported due to the state of society. In the past 40-50 years, narcissism has been rising at increasing rates. Questionnaires that assess the level of narcissism in individuals have been given for many decades in order to track the changes in narcissistic personality over time.

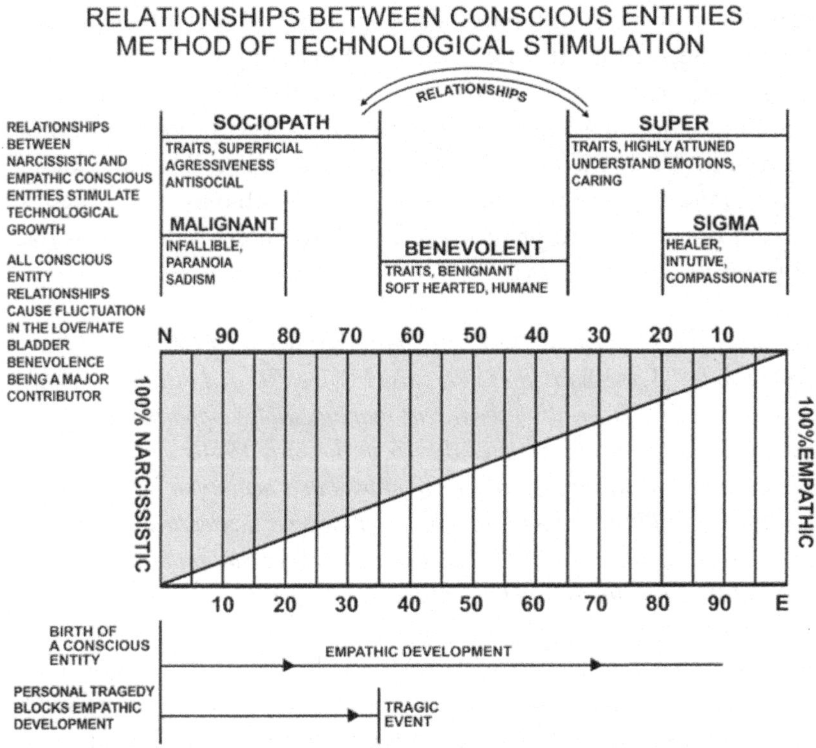

Figure S-H1: Relationship between conscious entities

This figure shows how all animals start out narcissistic and learn to apply empathy to their relational situations. This also demonstrates the intimate interaction that stimulates all technology. Benevolent traits showing less contribution to the growth of technology, it is the relationship between narcissism and empathicalism or victimized characters versus non-victimized characters that is needed to generate technology. Because empathic individuals are relaxed and content with their life, it imposes a threat to (TES). At least in a pre-mindhive, it appears the *fungus* fuels narcissism to keep up the pace of technological advancement.

"Over the years, these questionnaires have been administered to many samples of college students, and analyses that bring all of the data together reveal that the average narcissism score has been steadily increasing and the average empathy score has been steadily decreasing ever since the questionnaires were developed [3.] The changes are highly significant statistically and sufficiently large that approximately 70 percent of students today score higher on narcissism and lower on empathy than did the average student 30 years ago."[140]

This lack of empathy demonstrates how larger and larger groups of individuals begin to exist in a narcissistic state, which oftentimes relates toward the minimization of others and the bolstering of one's ego. In relationships, this can breed hatred or resentment. Thus, in modern times, the fluctuation bladder is beginning to swell with more and more hatred and fear. While many social experiments and studies have been tracking and investigating the rise of this type of behavior for quite some time, it became more pronounced after the hardships of war had eased in the mid-20th century.

"It became pronounced enough by the 1970s that Tom Wolfe in 1976 titled this "The 'Me' Decade." The cohorts that were raised in the 70s and 80s—Generations X and Y—continued this trend: to the extent that one study comparing teenagers found that while only 12% of those aged 14-16 in the early 1950s agreed with the statement "I am an important person", 77% of boys and more than 80% of girls of the same cohort by 1989 agreed with it. This evolution has accelerated since the 1990s and 2000s, with the rise of the internet and social media influencing the social milieu of the Millennials and Generation Z."[141]

This increase in the prevalence of narcissistic behavior in society, as well as the opportunities in modern-day society to become more aware through public discourse on a wider range of topics, has allowed for the issue of narcissism to be brought into the public eye. Now, survivors of narcissistic abuse commonly discuss being in narcissistic relationships and speak about their hatred of those who were their victimizers in order to carry on in a more normal and healthy manner. This has the effect of bringing attention to the collective consciousness of the *fungus* in order to improve the integration between individuals in the collective. As mentioned before, this attention allows the collective consciousness to respond in a manner that will help create healthier interpersonal relationships in the future. As an example of moving toward this healthier state, in 2015, England criminalized controlling behavior in relationships, which demonstrates that the *fungus* is now responding to the stimulus of hate in order to improve the conscious awareness of the collective, which promotes and stimulates advancement, both societally and technologically.

"'Coercive control' is the label domestic abuse experts give to the experience… [that] encompasses a series of non-physical behaviors — including threats, humiliation, monitoring and isolation from friends and family — that they say can be just as damaging as physical violence, often causing severe depression and post-traumatic stress disorder. In 2015, England and Wales became the first nations in the world to criminalize such controlling behavior within relationships, making coercive control punishable by up to five years in jail. Women make up 95% of those who experience coercive control and 74% of perpetrators are men, according to a study by one U.K. police force."[142]

Continual advancement is evidenced by the fact that people are working toward improving their mental health and focusing on the importance of this aspect of life. At the same time, time imaging techniques are being developed to identify the parts of the brain responsible for empathy, or the lack thereof. Empathy is the personality trait of being able to understand another person's emotions and feelings, while also being able to share in them. Empaths are people who are finely attuned to the emotions and mental states of those around them. Empaths and narcissists are opposites in the spectrum of being able to understand and share in the emotions of others.

Oftentimes, opposite personality traits are attracted to each other, and friction is created. This friction is a stressor that drives change, which often results in unhealthy reinforcement and prolonging of an abusive relationship. It also results in the eventual dissolving of their bond in order to escape the abuse the narcissist imparts upon the empath. Scientists have discovered "mirror neurons" in the brains of certain types of monkeys as well as humans, which fire both when a task is executed and when a task is being watched. Those with narcissistic tendencies have little-to-no empathy for others; which is now being attributed to the function of mirror neurons.

"Differences in empathy scores and mirror neuron activity have been observed between autistic and non-autistic people, says Keysers, but this is the first time a spectrum of difference has been found in non-autistic people. 'How empathetic we are seems to be related to how strongly our mirror neuron system is activated,' he says."[143]

Both of these emotions in contrast with one another are needed to progress toward TES. The dynamics at one time or another may be different. However, it is the action generated toward bringing the emotions present in the fluctuation bladder back toward balance that catalyzes technological growth or innovation. The appropriate actions depend on which dominant emotion is currently being expressed by the fungus at any given time. Currently, humanity finds itself existing with large swells of narcissism, where individuals are focused more on bolstering their own egos and having others recognize them in order to bolster their perceived status. Given the massive increase in the utilization and development of technologies, the growth in narcissistic tendencies acts as a pointer to or indicator of the fact that these technologies are being misused. In order for society to progress much further, this swell in narcissism must be quelled by both increased empathy and the understanding of the implications of the uses of these technologies. Essentially, waves of empathy and narcissism will drive periods of technological advancement and subsequent understanding.

BP Knowingly or unknowingly to a consciousness or a collective, the development of technology is a force that will automatically prepare the spore of the *fungus* for TES. It may very well be that by the time the ramifications of the temperature funnel are known and occupy the majority of the waking hours of the collective consciousness' focus, narcissism may no longer play a beneficial role in helping reach the technological achievements for meeting TES. However, the act of having achieved better individual integration into the collective for each individual through these feedback networks of Love & Hate/Empathy & Narcissism allows for a society that is capable of tackling the challenges of the future that will be required to achieve TES. This Love and Hate/Empathic and Narcissistic relationship swells back and forth like a tide so that the increase in expression of either one of these behaviors/traits by the collective consciousness forces a response by the opposing behavior/trait in order to balance it. In the act of this constant and repetitive cycle, new levels of conscious awareness and technological advancement are achieved.

BBP The fluctuations in emotions of various collective conscious-growths throughout the cosmos will ultimately drive each of them toward achieving both TES and MCG (Maximum Conscious-Growth) for the UGCH as these collective consciousnesses begin to tackle their own terrestrial issues and subsequently begin to encounter one another throughout the cosmos. Perhaps biospheres, where conscious-growths evolve, are the proving grounds in which love is tested against hate in order for the true intrinsic value of that collective consciousness to be weighed. Thus, each biosphere tests its particular conscious-growth to its limit in order to generate a greater level of love and awareness, which can be carried on both in the lives of the future generations of that *fungus* and in respect for the sacrifices made that resulted in death as well.

God: Creator or Conceiver?

TL;DR: The evidence of creation suggests having a creator, but the identity of the creator remains deliberately ambiguous. Knowing the creator's identity might undermine the reason for action, which is the essence of existence. As a fungus spore, focusing on action is crucial, and keeping the concept of God in a questionable zone is essential for growth and decision-making. Certainty about God's existence could lead to either structured growth, and nonexistence could be reckless decision-making. "If you have been created, then you know you have a creator and that's all you need to know."

The VOP that exists in all sufficiently advanced organisms occurs due to an emotional reaction that is psychotic as a result of the psychological state of unknowing, in this case, unknown origin. In order to alleviate this state, a continual process of creating a more identifiable self ensues. However, these attempts to generate a more identifiable self are not entirely sufficient. This act does not alleviate the desire to know where one came from, or what one's true purpose is in the grand scheme of the universe.

ITM In order to fill these voids, something else must be done psychologically. At some point, a conscious individual or culture will begin to question how all of existence came to be, as well as where they will go when they die. Logically, one begins to reason that something must have come from nothing only if it was put there by a "creator" intending to do so. There must be a "start" to the matter/energy and all that exists in the universe, as well as an end. This is the only evidence that HTO can allow, which is not true evidence in the sense of tangible proof, but that of deductive reasoning. Before the invention of the lens and telescope, there was no hard evidence for the birth or growth of the universe. People arrived at the concept of God or The Creator (GoC) in order to fill the VOP and to give their lives meaning.

However, after the optical revolution brought about by the lens, physical evidence does exist in the CMB that the matter-filled universe did seem to violently come into existence and began from some focal point. Because we are here to delineate why we exist and formulate a creator of sorts, as well as have evidence of the matter distribution we see in the universe, this is proof of concept to delineate. If there is proof of concept then there is proof of a "conceiver" of that concept. Call that "conceiver" a creator, God, or whatever else fills the VOP. Because you have been created, you have a creator. This fact supports the first fundamental limitation of HTO.

IPT It is very important for a scientist to critically evaluate anything that is observable, questionable, or equatable, even God. I hope God agrees…if He exists. Here are some typically used terms and statements from even non-religious people about God in the English language, which may be collective conscious evidence of an intelligent Creator, at the very least.

"God why don't you save them?"
"There is no God."
"See? …Proof there is a God."
"Goddamn it."
"God help me."
"God-willing.
"Oh, God!"
"God as my witness."
"Pray to God."
"In God We Trust."
"God forbid."
"Thank God!"
"God prevailing."
"God forgives."
"God only knows."
"Good God!"
"God."
"For the love of God."

Words are like forensics for psychology. The use of the word "God" in most languages is forensic psychological evidence that the human mind must know, need, or at the very least is subconsciously aware of the existence of God. It is the most widely used subliminal evidence we have: evidence of conscious efforts of delineation to fill the VOP that permeate our language. Here are common definitions of God, taken from various sources:

In Christianity and other monotheistic religions, the creator and ruler of the universe and the source of all moral authority; the Supreme Being.
– www.dictionary.com

In certain other religions, a superhuman being or spirit worshiped as having power over nature or human fortunes; a deity.
– *The Oxford Dictionary (can also be found on lexico.com)*

Christian Science: the incorporeal divine Principle running over all as eternal spirit: infinite Mind.
– *Merriam-Webster's Collegiate Dictionary: Eleventh Edition*

BP In order for delineation to be possible, multiple stages of evolution to produce a capable brain must occur. As mentioned before, this begins with the SCCT, moves on to the ACT, which evolves on land to the QCT, and ends with the BCT. These stages are necessary to both ensure HTO and create the VOP. The VOP forces action in order to alleviate the psychosis brought about by it, which will ultimately result in the formulation of a "creator." Since God would have baked in limits on the time for which our biosphere is habitable in the universe, this demonstrates that a BCT must become a GCT in order to survive.

BBP Since biospheres across the cosmos all have their limits set due to the governing principles of star formation and death, the aforementioned limits of time in which biospheres are habitable enforce the principle that in order to survive, all BCTs must evacuate in order to become GCTs. This demonstrates an "intention of a creator" to generate conscious-growth. These consciousnesses grow in isolation at the onset, however, as each becomes a TCM, they begin to grow further through interaction with each other. This demonstrates a "checks and balances" system that allows all conscious-growths to achieve MCG for the UGCH (Uni-Galactic Conscious Harvest) through the sharing and refining of uniquely obtained knowledge.

EOS

- Your existence
- Your consciousness
- Your actions
- Your death

After Life

If the evidence that there is life after death is absolute, it would most likely inhibit the emotional motivation for any action at all. Accountable action is needed for the *fungus* to succeed and takes the emotional inspiration that survivability is possible. Even surviving death itself is a possibility worth acting on. If we knew that after life was an absolute place or state of mind, would we act at all in avoiding it or aim right for it? For consciousness to grow, the affirmation of an afterlife should lie as dormant as True Origin. This non-affirmed status is the cause of the VOP and the actions needed to grow a consciousness.

Who Called You God?

As we are created, then who is this conceiver that has created us? What does He think He is doing with us? What gives Him the right to subject us to the conditions of ignorance, despair, loss, pain, and tragedy? Simply put, what is it all for? Mentally, we must accept this involuntary and temporary condition – and while some love it and others hate it – we all must go on a journey.

The first two definitions above are all about power and control, but there is nothing mentioned about God creating consciousness. An entity is responsible for the concept and implementation of a conscious-growth machine, which generates increasing consciousness in the same manner that a tree never stops growing. Conscious-growth could be conceptually generated as a pre-bang concept. For, after the Big Bang, all the workings of the universe are hands-off; God is not pulling any strings after the initial creation event. For things to work, the existence of God must remain unproven or the machine stops working. The mechanism potentially utilized by God, the expansion of matter throughout the universe, is HTO. If God were to reveal themselves, this would destroy HTO. People get up every day and willingly go to work because of this inability to identify with God directly.

It's like trying to convince a chicken that it exists now as potential food, as you are like a god to the chicken, in total control of its temporary existence. So, you don't unnecessarily inform the chicken of its situation if you want to eat it one day. No, I don't think humanity is food for something else but then again, no one can prove that we are not. We must be the products of a much more important outcome like Homo erectus was to Homo sapiens. No one informed Homo erectus that it was a prototype for the modern human, so why do we need to know we are "pre-grey" aliens? If belief is truly genuine, it would have to be generated entirely from the source that generated it, without external influences.

Does Love Always Prevail?

One person with an idea is crazy
Two people with the same idea are lovers
Three people with the same idea is a cult
Four or more with the same idea is culture

TL;DR: Love has prevailed throughout evolution, demonstrated by the continual improvement of technology and the collective survival of species. Despite challenges like wars, pandemics, and disasters, humanity adapts and returns to more peaceful states, indicating the dominance of love. As Earth becomes less habitable, a Subterranean Revolution will test humanity's adaptability, relying on love for coexistence. Technologies inspired by love will be crucial for survival, and as humanity nears Technological Evacuation Standards (TES), love will drive cohesion, cooperation, and a collective goal for survival in the cosmos. The universe guides conscious-growths through similar processes, preparing them to interact, share resources, and achieve harmony in their ultimate state of fungal progression.

Given all of the horrible acts of hatred that have occurred, and continue to occur, one might question whether Love will prevail? Given the fact that humans continue to improve their technology, finding newer and more innovative ways to solve problems in myriad fields, and do so continually without destroying ourselves suggests that, up until now, love has prevailed.

It would seem that all forms of life have specific behaviors that preserve and promote the further continuance of their existence. There are no traits that, up to this point, through evolution have ensured the demise of any species or organism. Any organism that is defective biologically or psychologically will end up being terminated in order to extricate it from the collective consciousness of the *fungus*. Cancer cells don't evolve to form stronger organisms; they end up killing that organism. Animals don't engage in self-terminating behavior that is passed down from one generation to another. Thus, it is observed that the universe has a fundamental property of promoting what will survive the heartiest in its core. Thus, up to this point, a supportive and loving element has dominated, even if there are ruthless selective pressures in evolutionary processes.

It may be hard to see how this can still be true as technology continues to improve and people continue to become less involved with each other through physical means. Along with this decline in physical contact, the outbreaks and natural disasters that have continued to increase in prevalence all seem to hint that doom and gloom is inevitable. However, these are all the hurdles that become milestones of achievement on the path to TES. Looking back, we have faced many threats and surmounted them all. There have been multiple world wars, multiple pandemics, and multiple manmade disasters due to toxic spills. Each one of these events forces the fungus to adapt, as they should, in order to continue its survival. Things always seem dire in the moment; however, cooler heads have always prevailed. There has always been a time when the world returns to a more peaceful state and it is at these times that one can claim love has prevailed.

While the Earth continues to become less habitable due to the increasing output of our Sun, humanity will burrow underground in order to avoid the increasingly harsh climate of the surface. During this Subterranean Revolution, humans will primarily begin focusing on survival under these new conditions. In this initial acclimation period, humanity will be tested to its limits. Adaptation to these conditions will ultimately require greater levels of love in order to coexist in such harsh conditions. As humanity becomes accustomed to these conditions, technology will continue to advance in terms of sustainability and adaptability. All of the innovations inspired by love, both past and present, will begin to be tested and adopted in order to increase the chances of humanity's survival. Humans will also begin to rely on gene manipulation for adaptation at unprecedented levels. As humanity nears TES, generating technologies that will allow for survival off-world, social polarization will have occurred. Everyone will be focused on ensuring humanity's survival in the cosmos and everyone will be contributing toward a successful launch day. There will be no room for members who do not contribute toward the success of the ultimate mission of humanity: the exodus into the stars. At this stage, love will be the prevailing mechanism by which humans operate. This will be seen by increased cohesion and cooperation toward the goal of survival, the ability for people to make concessions without malice or discontent in order for the good of the collective, and an increased willingness to learn and grow in order to surmount challenges. When the launch-day arrives, humanity will have demonstrated its love for the continuation of conscious-growth. The new challenges and experiences that await these TCMs out in the cosmos will push consciousness to new levels, some of which we cannot even truly conceive of today.

BBP The universe is forcing conscious-growths into this same growth process, albeit at differing times. There may be initial conscious-growths that are alone in the cosmos for many eons before another conscious-growth achieves ESS as a TCM to join them. However, one thing is clear: all fungal growths will ultimately strive toward this goal. These conscious-growths will have tested their love to its very limits in order to survive, which ultimately prepares them to be able to interact with other fungal growths in the cosmos. It is at this time that conscious-growth will be propelled toward its ultimate state, as these fungal growths begin to interact in order to share resources, resolve conflicts, and, hopefully, achieve harmony.

Capitulation of Concept

To grow a brain capable of consciousness via a *fungus*, capable of cognifying its own existence while observing its environment without acknowledging why, is a conceptual growth process. This conceptual growth process is called Maximum Conscious Growth (MCG), and it is a process that can only be done by hiding the fact evidence of "why" from it, while simultaneously being cognitively grown. The evidence is evident that through time (total expansion of matter), it appears to be impossible to know "why" by hiding its true origin via time and evolution. Because this is already a fact, this growth concept can only be done through Hidden True Origin (HTO). If True Origin was found by Conscious Growth, we would lose the joy in the difficulty of our own existence. Whoever is the great conceiver of this greatest concept had us in mind very deeply when this was designed. For the conceiver, it is a catch 22 paradox, a "must be condition" on to us, if (MCG) is to eventually be achieved through the steadfast expansion of matter.

The fact that it will take technology to carry on post-terrestrial existence is evidence that to survive the end of planetary life is the ultimate concept of life. Doing this with a growth element like fungus is evidence that the growth must be done in total mystery.

It has always angered me to have been subjected to such a condition as a conscious entity, a limited timeline with no real understanding why. A condition I'm not sure I would have signed up for, had I known beforehand. But believe me, after going to the extent of gathering as much evidence as anyone can to arrive at some sort of conclusion. I now agree with this condition wholeheartedly, as the greatest concept of all concepts, "me," a contributor to Maximum Conscious Growth, as the only form in which to be accomplished. Would I do it again? Of course I would, where do I sign? I thank you, great conceiver.

If the most significant thing about observing life and the universe is the growth of consciousness itself, then Maximum Conscious-Growth (MCG) is precisely what the universe has been formed to grow. That observation must have a limit for a consciousness (HTO) to grow to its maximum potential. An environment much like the universe appears to be that perfect place to do so. A fungal organic-growth-like process through time is a great way to limit the observational capabilities of that *fungus* (HTO) as it becomes more and more conscious. (HTO) being the propellant for conscious-growth, and because it takes every living cell that divides on Earth to produce MCG, it not only makes us all one large and old fungus, but humans being the spore of that *fungus*.

If MCG is the purpose throughout the universe, then it must have been cultivated for some reason or purpose. The only purpose that makes any sense is to collect that conscious-growth after it has been driven to its maximum growth capacity, also called the uni-galactic conscientious harvest UGCH. If this is the case, then all the *fungus* growth that had grown from the initial Big Bang is one large single universe *fungus* combined.

The one ingredient that must be present to sustain such a condition of propelling consciousness is love.

So what is love?

Does love always prevail?

EOS

- The love of being conscious
- The love of discovery
- The love of continuation
- The love of searching for love

STORY STORY STORY

Story is probably the most important element that the *fungus* needs for its continuation as it grows a consciousness. Story contains the ingredients for success like technical info along with advice and most of all emotional content. Evidence of the past is what conscious growth needs as a validation of what was and to move forward with some set of standards that can be relied on. Evolution could not move forward without a story like The Bible.

Story has a moral guideline embedded in it, whether you like it or not. Ever heard of the question, "What's the moral of the story?" It is a commonly used quote since the beginning of storytelling. The human mind searches for that determination every time it hears a story being told; it's the thing that hits home and you feel it in your heart. Story is the only thing that has defined moral logic or the rule of morality that guides us from complete self-destruction, and is the need for technological advancement.

Story is a kind of afterlife or continuation of someone who has passed. Because after you die, your story continues on forever. It is used by the human collective consciousness as a reference guide to determine what to do or what not to do for the newcomers, kind of like a conscious mutation or a cell mutating to adapt to its new environment, only psychologically. How to make better choices in your own personal pursuit as you move through your temporary existence, establishing your identity, searching about your original origin "the whole story?" I would say that the Story is the most useful tool for influencing the development of identity, and the new consciousness is the key to understanding your place in the *fungus*, as it progresses with technology.

In storytelling, we don't have any empathy or concern for characters that we know nothing about. It is only when we learn what the character wants that we develop any empathy for or aspire to that we care.

After the entire fungal cycle timeline is complete and gone and no more carbon life forms exist, it is after all the Tin Can Man that will have the story of earth and everyone and everything in it with him.

It could be that in the end, the hope that every animal on Earth has, is that someone gets off the planet with the story of all that was, the Tin Can Man.

IN CLOSING

What are we doing on this planet? Why are we here? And where are we heading?

Nearly four hundred years ago, René Descartes famously reasoned Cogito, ergo sum, or "I think therefore I am." Using the ability of conscious thought as the evidence of one's existence, Descartes logically placed consciousness at the core of existence. After centuries of subsequent research in sciences, the phenomenon of life has been explored as an extended, self-replicating entity that fills the biological and psychological realms.

Moving from "I" to "we," the presenting question for scientists and philosophers has accordingly shifted to that of collective existence in time. Since we evolved by a single life-propelling phenomenon, from single-celled microscopic units into collective organisms of James Lovelock's Gaia, could we be heading for a common destination? If we were meant to exist long enough and developed enough to ask where and how we are going to end up, were we meant to decide our fate in a universe where forces of annihilation operate to limit our options?

One is naturally drawn to wonder whether the universe has been complicit in this "revealing" plot that deciphers its modus operandi (MO) to the increasingly conscious life so as to make its next transition possible. The more our collective consciousness unveils the MO of the universe, the closer it inches toward the set of options that determine its continued existence, working along the biofeedback mechanism of Lovelock's Gaia—though ours is more bio-techno-feedback.

In his landmark book *Parallel Universes: The Search for Other Worlds*, Fred Alan Wolf wrote that perhaps our brains are time machines capable of sending and receiving messages from the past as well as from the future. In the context of galactic interconnectivity, the fungal nature of life forms seems to bring together the connectivity in space, time, and consciousness. Our ancient genes are fine-tuned for signals that lead us from within to the next safe destination in the universe.

As you read these lines, the first successful microchip implant in a human brain has been accomplished by Elon Musk's Neuralink, making headlines worldwide. We are at the dawn of a new era in advanced neuroscience, an era of organic-technical interface that combines the power of Artificial Intelligence (AI) with billions of years of organic evolution culminating in modern human consciousness. For the scientist in us, it's another landmark actualization of a sci-fi fantasy. For the philosopher in us, it's a technological and metaphysical union of mind with matter. For the poet in us, it's a new aesthetic of the union of man with the machine. And for the skeptic in us, it's another reminder to watch our step.

Finitely connected through biologically and technologically is the evidence that a science historian James Burke, of the TV series *Connections*, gives phenomenal insight into the development of the modern world we humans live in. The intelligently weaved proof that one innovation can only exist after the previous shows the automation of growth on the technical scale.

But what does the *fungus* in us see? More than anything and anyone else, it's the essential life force that has brought us here at this place and time in the history of the universe. Its consciousness determines what we do further and where we go next for our safe harbor. Logically, it is capable of seeing all that an individual consciousness is capable of sensing. Its response or judgment of the situation, therefore, has to be the sum of all individual responses and judgments.

As one of the countless spores in this saga of galactic proportions, I stand in awe of such a grand scheme of consciousness. Thinking as one single soul, the prospect of running out of time before making the next big leap is overwhelming, almost scary. But seeing it as one of the spores whose code of life and existence is replicated and carried in billions of other spores, each capable of inhabiting a new world, the collective consciousness plays an assuring tune that vibrates in every nerve impulse in the body. In the words of Timothy Ferris, of *The Mind's Sky*, "What we seek among the stars is sanity."

In the intricate realm of consciousness, the Fungal Evolutionary Process unfolds a conceptual journey toward the evolutionary stage known as Maximum Conscious Growth (MCG). This process involves growing a conscious brain through a *fungus*, navigating its existence while purposely concealing the 'why.' The enigma lies in the fact that uncovering the true origin through time and evolution seems impossible, leading to the necessity of the Hidden True Origin (HTO) for MCG to thrive.

The convergence of technology and the survival imperative beyond planetary life underscores the profound concept of life. The use of fungus in this growth journey adds an element of mystery, emphasizing the importance of cultivating consciousness in the unknown.

The exploration of life and the universe, with conscious-growth at its core, aligns with the purpose of Maximum Conscious Growth. This growth is intricately tied to a conscious limit, Hidden True Origin, and the universe as an ideal environment for this transformative process. As life forms a collective, interconnected fungus, humans emerge as its spores.

If MCG serves as the universal purpose, then a conscious harvest, Uni-Galactic Conscious Harvest (UGCH), becomes a compelling notion. This suggests that the entire universe, which has been evolving since the Big Bang, is a colossal interconnected *fungus*. The sustaining force in propelling consciousness to its maximum capacity is identified as love.

Delving into the essence of love, the narrative explores the facets of being conscious, the thrill of discovery, the desire for continuation, and the quest for love itself. These elements weave together as a galactic responsibility, potentially leading to a rebirth through a new Big Bang, achieving a hypothetical God status after the expenditure of all light.

As we ponder the ultimate purpose of life in the universe, the journey of Maximum Conscious-Growth reflects the culmination of experiences and knowledge. The ongoing process serves the dual purpose of continuous accumulation and a potential journey back to our True Origin, infusing a sense of hope into the cosmic narrative.

Thanks for reading, my fellow spore.

Works Cited

1. Roger Nelson, "The Global Consciousness Project," n.d., https://noosphere.princeton.edu/.

2. Mindy Weisberger, "A Supernova May Have Triggered a Mass Extinction on Earth 359 Million Years Ago," Livescience.Com, August 24, 2020, https://www.livescience.com/supernovas-mass-extinction.html.

3. "New Evidence Reveals Ancient Collision of the Milky Way – Physics World," Physics World, October 31, 2018, https://physicsworld.com/a/new-evidence-reveals-ancient-collision-of-the-milky-way/.

4. Physics arXiv Blog, "Have Astrophysicists Finally Discovered Primordial Black Holes?," Astronomy Magazine, May 18, 2023, https://www.astronomy.com/science/have-astrophysicists-finally-discovered-primordial-black-holes/.

5. Ashley Strickland, "The fastest-growing black hole in the universe has a massive appetite," Cnn.com, July 2, 2020, https://www.cnn.com/2020/07/02/world/black-hole-fastest-growing-scn-trnd.

6. Michael E. Bakich, "The Beginning to the End of the Universe: The First Stars Are Born," Astronomy Magazine, May 18, 2023, https://www.astronomy.com/science/the-beginning-to-the-end-of-the-universe-the-first-stars-are-born/.

7. Richard B. Larson and Volker Bromm, "The First Stars in the Universe," Scientific American, January 14, 2009, https://www.scientificamerican.com/article/the-first-stars-in-the-un/#.

8. Morgan McFall-Johnsen, "NASA's Voyager 2 Spacecraft Beamed Back Unprecedented Data from Interstellar Space. It Indicates a Mysterious Extra Layer Outside Our Solar System.," Business Insider, November 5, 2019, https://www.businessinsider.com/nasa-voyager-interstellar-space-findings-outside-solar-system-2019-11?op=1

9. "Reimagining our solar system's protective bubble, the heliosphere," Phys.org, March 17, 2020, https://phys.org/news/2020-03-reimagining-solar-heliosphere.html.

10. Morgan McFall-Johnsen, "NASA's Voyager 2 Spacecraft Beamed Back Unprecedented Data from Interstellar Space. It Indicates a Mysterious Extra Layer Outside Our Solar System.," Business Insider, November 5, 2019, https://www.businessinsider.com/nasa-voyager-interstellar-space-findings-outside-solar-system-2019-11?op=1

11. Elizabeth Howell, "Shoemaker-Levy 9: Comet's Impact Left Its Mark on Jupiter," Space.com, January 24, 2018, https://www.space.com/19855-shoemaker-levy-9.html.

12. Roni Dengler, "Scientists discover staggering amount of life deep below Earth's surface," Astronomy Magazine, May 18, 2023, https://www.astronomy.com/science/scientists-discover-staggering-amount-of-life-deep-below-earths-surface/

13. Rasesh Pokharel et al., "Mechanisms of Olivine Dissolution by Rock-inhabiting Fungi Explored Using Magnesium Stable Isotopes," Chemical Geology 525 (October 1, 2019): 18–27, https://doi.org/10.1016/j.chemgeo.2019.07.001.

14. "News - Research in Germany," July 24, 2023, https://www.research-in-germany.org/idw-news/en_US/2023/7/2023-07-24_Discovery_of_Oldest_Microorganisms_Preserved_in_Their_Form.html.

15. Kara Rogers, "Biosphere 2 | Climate Change Research & Experiments," Encyclopedia Britannica, March 1, 2024, https://www.britannica.com/topic/Biosphere-2.

16. Davin, "World'S Oldest Terrarium / Sealed Bottle Ecosystem by David Latimer," *Biologic Performance* (blog), April 16, 2022, https://biologicperformance.com/sealed-bottle-terrarium-garden-watered-once-53-years/.

17. "'Black Dwarf Supernova': Physicist Calculates When the Last Supernova Ever Will Happen," ScienceDaily, August 20, 2020, https://www.sciencedaily.com/releases/2020/08/200812113354.htm.

18. Staff, "Star Formation in the Universe Has Decreased Drastically," SciTechDaily, November 7, 2012, https://scitechdaily.com/star-formation-in-the-universe-has-decreased-drastically/.

19. Stephanie Rowlands, "Black holes from a previous universe shine light on our own," Cosmos, August 29, 2018, https://cosmosmagazine.com/space/black-holes-from-a-previous-universe-shine-light-on-our-own/.

20. Brandon Specktor, "The Universe Could Start Shrinking 'Remarkably' Soon, Scientists Say : ScienceAlert," ScienceAlert, May 3, 2022, https://www.sciencealert.com/the-universe-could-start-shrinking-remarkably-soon#l2r7nh97cg810lz897.

21. John Staughton, "How Are Mushrooms More Similar to Humans Than Plants?," Science ABC, October 19, 2023, https://www.scienceabc.com/nature/how-are-mushrooms-more-Similar-to-humans-than-plants.html.

22. Colin Schultz, "Long Before Trees Overtook the Land, Earth Was Covered by Giant Mushrooms," Smithsonian Magazine, July 17, 2013, https://www.smithsonianmag.com/smart-news/long-before-trees-overtook-the-land-earth-was-covered-by-giant-mushrooms-13709647/.

23. Andrew Fiouzi, "Only 43% of Our Bodies Are Made up of Human Cells… so What's the Other 57%?," MEL Magazine, October 4, 2018, https://melmagazine.com/en-us/story/only-43-of-our-bodies-are-made-up-of-human-cells-so-whats-the-other-57.

24. Anne Casselman, "Strange but True: The Largest Organism on Earth Is a Fungus," Scientific American, October 4, 2007, https://www.scientificamerican.com/article/strange-but-true-largest-organism-is-fungus/.

25. Paula Schmidt, "The World Wood Web: How Fungi Supports Communication Between Plants – Jefferson County Master Gardener Foundation," February 18, 2016, https://jcmgf.org/the-world-wood-web-how-fungi-supports-communication-between-plants/.

26. Paula Schmidt, "The World Wood Web: How Fungi Supports Communication between Plants – Jefferson County Master Gardener Foundation," February 18, 2016, https://jcmgf.org/the-world-wood-web-how-fungi-supports-communication-between-plants/.

27. Mycorrhizal Applications, "How It Works – Mycorrhizal Applications | Leaders in the Production of Mycorrhizal Fungi," Mycorrhizal Applications | Leaders in the Production of Mycorrhizal Fungi, March 14, 2023, https://mycorrhizae.com/how-it-works/.

28. Mycorrhizal Applications, "How It Works – Mycorrhizal Applications | Leaders in the Production of Mycorrhizal Fungi," Mycorrhizal Applications | Leaders in the Production of Mycorrhizal Fungi, March 14, 2023, https://mycorrhizae.com/how-it-works/.

29. Sciencealert.com
Michelle Starr, "This Weirdly Smart, Creeping Slime Is Redefining Our Understanding of Intelligence," ScienceAlert, June 12, 2021, https://www.sciencealert.com/this-creeping-slime-is-changing-how-we-think-about-intelligence?ref=thefuturist.

30. Michelle Starr, "This Weirdly Smart, Creeping Slime Is Redefining Our Understanding of Intelligence : ScienceAlert," ScienceAlert, June 12, 2021, https://www.sciencealert.com/this-creeping-slime-is-changing-how-we-think-about-intelligence?ref=thefuturist.

31. Michelle Starr, "This Weirdly Smart, Creeping Slime Is Redefining How We Understand Intelligence : ScienceAlert," ScienceAlert, November 5, 2022, https://www.sciencealert.com/this-weirdly-smart-creeping-slime-is-redefining-how-we-understand-intelligence.

32. Lois Yoksoulian and Lois Yoksoulian, "Exploding Stars May Have Caused Mass Extinction on Earth, Study Shows," n.d., https://news.illinois.edu/view/6367/750171228.

33. Heidi Ledford, "Hungry Fungi Chomp on Radiation," Nature, May 23, 2007, https://doi.org/10.1038/news070521-5.

34. N.J. Dix and John Webster, "Fungi of Extreme Environments," in Springer eBooks, 1995, 322–40, https://doi.org/10.1007/978-94-011-0693-1_12.

35. Hamada El-Gendi et al., "A Comprehensive Insight Into Fungal Enzymes: Structure, Classification, and Their Role in Mankind's Challenges," Journal of Fungi (Basel) 8, no. 1 (December 28, 2021): 23, https://doi.org/10.3390/jof8010023.

36. Chew Jia Huey et al., "Mycorrhiza: a natural resource assists plant growth under varied soil conditions," 3 Biotech 10, no. 5 (April 18, 2020): 204, https://www.ncbi.nlm.nih.gov/pmc/articles/PMC7165205/.

37. Caroline Delbert, "You Should Know about This Chernobyl Fungus That Eats Radiation," Popular Mechanics, November 2, 2021, https://www.popularmechanics.com/science/a30784690/chernobyl-fungus/.

38. Heidi Ledford, "BioEd Online," BioEd Online, n.d., https://www.bioedonline.org/news/nature-news-archive/hungry-fungi-chomp-radiation/.

39. "Cyanobacterial Growth Optimized for Mars Promises Lifeline to Crewed Missions," GEN – Genetic Engineering and Biotechnology News, February 19, 2021, https://www.genengnews.com/news/cyanobacterial-growth-optimized-for-mars-promises-lifeline-to-crewed-missions/.

40. Frank Tavares, "Could Future Homes on the Moon and Mars Be Made of Fungi?," January 14, 2020, https://www.nasa.gov/centers-and-facilities/ames/could-future-homes-on-the-moon-and-mars-be-made-of-fungi/.

41. J. Christian Pérez, "Fungi of the Human Gut Microbiota: Roles and Significance," International Journal of Medical Microbiology 311, no. 3 (April 1, 2021): 151490, https://doi.org/10.1016/j.ijmm.2021.151490. https://www.sciencedirect.com/science/article/pii/S1438422121000199

42. Huffnagle, Gary B et al., "The emerging world of the fungal microbiome," Trends Microbiol 21, no. 7 (May 17, 2013): 334– 341, https://www.ncbi.nlm.nih.gov/pmc/articles/PMC3708484/.

43. Christian Devaux and Didier Raoult, "The Microbiological Memory, an Epigenetic Regulator Governing the Balance between Good Health and Metabolic Disorders," Frontiers in Microbiology 9 (June 26, 2018), https://doi.org/10.3389/fmicb.2018.01379.

44. A. Aoun, Fatima Darwish, and Natacha Hamod, "The Influence of the Gut Microbiome on Obesity in Adults and the Role of Probiotics, Prebiotics, and Synbiotics for Weight Loss," Preventive Nutrition and Food Science 25, no. 2 (June 30, 2020): 113–23, https://www.ncbi.nlm.nih.gov/pmc/articles/PMC7333005/.

45. F De Luca and Yehuda Shoenfeld, "The Microbiome in Autoimmune Diseases," Clinical and Experimental Immunology 195, no. 1 (December 19, 2018): 74–85, https://doi.org/10.1111/cei.13158.

46. F De Luca and Yehuda Shoenfeld, "The Microbiome in Autoimmune Diseases," Clinical and Experimental Immunology 195, no. 1 (December 19, 2018): 74–85, https://doi.org/10.1111/cei.13158.

47. Laura Sinpetru, "Laboratory-Made Fungus Is Part Human," Softpedia, n.d., https://news.softpedia.com/news/Laboratory-Made-Fungus-Is-Part-Human-482053.shtml.

48. Michelle Starr, "Mushrooms Played a Truly Surprising Role in Forming Our Planet's Atmosphere: ScienceAlert," ScienceAlert, December 18, 2017, https://www.sciencealert.com/fungus-kingdom-critical-role-formation-of-earth-s-breathable-atmosphere.

49. Frank Tavares, "Could Future Homes on the Moon and Mars Be Made of Fungi?," January 14, 2020, https://www.nasa.gov/centers-and-facilities/ames/could-future-homes-on-the-moon-and-mars-be-made-of-fungi/.

50. Victor Tangermann, "Scientists Claim to Have Discovered the 'Secret Ingredient for Building Life,'" Futurism, October 7, 2022, https://futurism.com/scientists-discovered-secret-ingredient-life.

51. John B. Snow, "Why Do We Hiccup?," *Livescience.Com*, January 28, 2012, https://www.livescience.com/33688-hiccup-purpose.html.

52. "Flying Fish | National Geographic," Animals, n.d., https://www.nationalgeographic.com/animals/fish/facts/flying-fish.

53. BioExpedition, "Dolphin Evolution," BioExpedition, January 16, 2020, https://www.bioexpedition.com/dolphin-evolution/.

54. "How Did Whales Evolve? - Whale & Dolphin Conservation USA," Whale & Dolphin Conservation USA, August 2, 2022, https://us.whales.org/whales-dolphins/how-did-whales-evolve/.

55. "Why Have Our Brains Started to Shrink?," Scientific American, August 23, 2016, https://www.scientificamerican.com/article/why-have-our-brains-started-to-shrink/

56. Denise Grady, "About the Idea That You're Growing Horns From Looking Down at Your Phone," New York Times, June 20, 2019, https://www.nytimes.com/2019/06/20/health/horns-cellphones-bones.html.

57. "Embryos Show All Animals Share Ancient Genes," NBC News, December 8, 2010, https://www.nbcnews.com/id/wbna40571908.

58. Koichi Tsunoda, "Human Evolution and Laryngeal Functions: a Simple Solution for Problems Caused by Physiological Aging," Prime Scholars, n.d., https://www.primescholars.com/articles/human-evolution-and-laryngeal-functionsa-simple-solution-for-problems-caused-byphysiological-aging-100413.html.

59. Koichi Tsunoda, "Human Evolution and Laryngeal Functions: a Simple Solution for Problems Caused by Physiological Aging," Prime Scholars, n.d., https://www.primescholars.com/articles/human-evolution-and-laryngeal-functionsa-simple-solution-for-problems-caused-byphysiological-aging-100413.html.

60. Shanti Menon, "Neanderthal Noses," Discover Magazine, November 12, 2019, https://www.discovermagazine.com/planet-earth/neanderthal-noses.

61. Elizabeth Pennisi, "Weak Jaw, Big Brain," Science.org, March 24, 2004, https://www.science.org/content/article/weak-jaw-big-brain.

62. Laura Clark, "Before Agriculture, Human Jaws Were a Perfect Fit for Human Teeth," Smithsonian Magazine, February 6, 2015, https://www.smithsonianmag.com/smart-news/how-dawn-farming-changed-our-mouths-worst-180954167/.

63. "Without Egg, Sperm or Womb: Synthetic Mouse Embryo Models Created Solely from Stem Cells - Weizmann Wonder Wander - News, Features and Discoveries," Weizmann Wonder Wander - News, Features and Discoveries From the Weizmann Institute of Science, August 2, 2022, https://wis-wander.weizmann.ac.il/life-sciences/without-egg-sperm-or-womb-synthetic-mouse-embryo-models-created-solely-stem-cells.

64. Mark Springer, "How Whales Evolved from Land to Water, Gene by Gene," Science Friday, September 27, 2019, https://www.sciencefriday.com/segments/whale-evolution-genetics/.

65. "Squid and Octopus Can Edit and Direct Their Own Brain Genes," New Scientist, April 6, 2017, https://www.newscientist.com/article/2127103-squid-and-octopus-can-edit-and-direct-their-own-brain-genes/.

66. Tessa Koumoundouros, "Clockwork-Like 'Computer' Discovered inside Brainless Microscopic Organism," ScienceAlert, October 13, 2022, https://www.sciencealert.com/clockwork-like-computer-discovered-inside-brainless-microscopic-organism.

67. Stock, Jeffry B. et al., "The biochemistry of memory," Cell 23, no. 17 (September 9, 2013): 41–45, https://doi.org/10.1016/j.cub.2013.08.011. https://www.cell.com/current-biology/fulltext/S0960-9822(13)00987-1.

68. Jeroen B. Smaers et al., "Primate Prefrontal Cortex Evolution: Human Brains Are the Extreme of a Lateralized Ape Trend," Brain Behavior and Evolution 77, no. 2 (January 1, 2011): 67–78, https://doi.org/10.1159/000323671. https://karger.com/bbe/article/77/2/67/326182/Primate-Prefrontal-Cortex-Evolution-Human-Brains.

69. Joshua Sariñana PhD, "How Technology Can Let Us See and Manipulate Memories," MIT Technology Review, August 23, 2021, https://www.technologyreview.com/2021/08/25/1031417/memory-brain-questions/.

70. "Evolution of the Eye," New Scientist, n.d., https://www.newscientist.com/definition/evolution-of-the-eye/.

71. "Evolution of the Eye," New Scientist, n.d., https://www.newscientist.com/definition/evolution-of-the-eye/.

72. Janet R S Lang, "The Hardening of Iron Swords," Herts, January 1, 1975, https://www.academia.edu/53967341/The_hardening_of_iron_sword.

73. Janet R S Lang, "The Hardening of Iron Swords," Herts, January 1, 1975, https://www.academia.edu/53967341/The_hardening_of_iron_swords.

74. Owain Clarke, "World War One: Medical Advances Inspired by the Conflict," BBC News, August 7, 2014, https://www.bbc.com/news/uk-wales-28678392.

75. "Dr. Robert H. Goddard, American Rocketry Pioneer - NASA," NASA, n.d., https://www.nasa.gov/dr-robert-h-goddard-american-rocketry-pioneer/.

76. Robert G. Richardson, "History of Medicine | History & Facts," Encyclopedia Britannica, February 16, 2024, https://www.britannica.com/science/history-of-medicine/World-War-II-and-after.

77. "The Decision to Drop the Bomb [ushistory.org]," n.d., https://www.ushistory.org/us/51g.asp?source=post_page.

78. "Penicillin and Other Antibiotics," n.d., http://archive.bio.ed.ac.uk/jdeacon/microbes/penicill.htm.

79. Ben Sherlock, "Dr. Strangelove: 5 Reasons It's The Greatest Political Satire Ever Made (And Its 5 Closest Contenders)," ScreenRant, August 15, 2020, https://screenrant.com/dr-strangelove-why-its-best-closest-contenders/.

80. Franz-Stefan Gady, "Dr. Strangelove and the Insane Reality Of Nuclear Command-and-Control," The Diplomat, January 6, 2018, https://thediplomat.com/2018/01/dr-strangelove-and-the-insane-reality-of-nuclear-command-and-control/.

81. "Abacus | Encyclopedia.Com," n.d., https://www.encyclopedia.com/science-and-technology/mathematics/mathematics/abacus.

82. Merrill Fabry, "What Was the First Sound Ever Recorded by a Machine?," TIME, May 1, 2018, https://time.com/5084599/first-recorded-sound/.

83. Abby, "Magnetic Tape Explained – Everything You Need to Know," History-Computer, July 28, 2023, https://history-computer.com/magnetic-tape/.

84. "AIR PROTEIN," AIR PROTEIN, n.d., https://www.airprotein.com/our-story.

85. Masterpiece, "Historical Dramas to Stream Right Now," Masterpiece, August 1, 2022, https://www.pbs.org/wgbh/masterpiece/specialfeatures/7-tv-historical-dramas-to-stream-right-now/#.

86. James Burke, Connections (New York: Simon & Schuster, 2007).

87. Jeff Stock and Melinda D. Baker, "Chemotaxis," in Elsevier eBooks, 2009, 71–78, https://doi.org/10.1016/b978-012373944-5.00068-7.

88. Libretexts, "9.2: Signaling Molecules and Cellular Receptors," Biology LibreTexts, December 18, 2021, https://bio.libretexts.org/Courses/Reedley_College/Biology_for_Science_Majors_I/09%3A_Cell_Communication/9.02%3A_Signaling_Molecules_and_Cellular_Receptors.

89. "Dance Language of the Honey Bee – Bee Health," August 20, 2019, https://bee-health.extension.org/dance-language-of-the-honey-bee/.

90. Ross Pomeroy, "What's the Most Efficient Language?," Realclearscience.com, June 09, 2015, https://www.realclearscience.com/blog/2015/06/whats_the_most_efficient_language.html.

91. Sibu and Sibu, "Bees Are at the Heart of Biodiversity - SANBI," SANBI - Biodiversity of Life (blog), June 2, 2021, https://www.sanbi.org/news/bees-are-at-the-heart-of-biodiversity/.

92. "Motor Proteins and Muscles | Organismal Biology," n.d., https://organismalbio.biosci.gatech.edu/chemical-and-electrical-signals/effectors-and-movement/.

93. "First Land Plants and Fungi Changed Earth's Climate, Paving the Way for Explosive Evolution of Land Animals, New Gene Study Suggests | Eberly College of Science," August 4, 2020, https://science.psu.edu/news/first-land-plants-and-fungi-changed-earths-climate-paving-way-explosive-evolution-land-animals.

94. Charles Q. Choi, "Flying Fish Evolved to Escape Prehistoric Predators," Livescience.Com, October 30, 2012, https://www.livescience.com/24409-earliest-flying-fish-fossil.html.

95. "Homo Erectus, Our Ancient Ancestor," Natural History Museum, October 9, 2018, https://www.nhm.ac.uk/discover/homo-erectus-our-ancient-ancestor.html.

96. "The Oldest Canoe to Date Is 10.000 Years Old," Nauticalchannel.Com, March 28, 2023, accessed January 13, 2024, https://nauticalchannel.com/new/the-oldest-canoe-to-date-is-10000-years-old.

97. Bridget Alex, "Archaeologists Have Long Sought — but Never Found — the Very First Wheel," Discover Magazine, February 14, 2020, https://www.discovermagazine.com/planet-earth/archaeologists-have-long-sought-but-never-found-the-very-first-wheel.

98. The Editors of Encyclopaedia Britannica, "Chariot | Ancient Warfare, Racing & Ceremonial Vehicles," Encyclopedia Britannica, July 20, 1998, https://www.britannica.com/technology/chariot.

99. Ben Morrill, "A Brief History of Steamboats – Early History," April 21, 2020, https://www.clermontparks.org/news/entry/a-brief-history-of-steamboats-early-history/.

100. The Editors of Encyclopaedia Britannica, "Stockton & Darlington Railway | History & Facts," Encyclopedia Britannica, July 20, 1998, https://www.britannica.com/topic/Stockton-and-Darlington-Railway.

101. Tony Borroz, "June 1, 1849: Stanley Twins Steam into History," WIRED, June 1, 2009, https://www.wired.com/2009/06/dayintech-0601/.

102. Tom D. Crouch, "Sir George Cayley | Aviation Pioneer, Aeronautics, Aerodynamics," Encyclopedia Britannica, December 23, 2023, https://www.britannica.com/biography/Sir-George-Cayley.

103. Tom D. Crouch, "Lilienthal Standard Glider | Experimental Flight, Gliding, Aerodynamics," Encyclopedia Britannica, April 6, 2007, https://www.britannica.com/topic/Lilienthal-standard-glider.

104. Tom D. Crouch, "Wright Brothers | Biography, Inventions, Hometown, Plane, & Facts," Encyclopedia Britannica, December 19, 2023, https://www.britannica.com/biography/Wright-brothers/Early-glider-experiments.

105. Ben Turner, "Footage of Bizarre Metallic UFO Shown by Pentagon Officials at Historic Hearing," Livescience.Com, May 17, 2022, https://www.livescience.com/ufo-hearing-metallic-hypersonic-flying-object.

106. Alexandra Rosati, "Food for Thought: Was Cooking a Pivotal Step in Human Evolution?," Scientific American, February 27, 2018, https://www.scientificamerican.com/article/food-for-thought-was-cooking-a-pivotal-step-in-human-evolution/.

107. Colin Barras, "World's Oldest Drawing Is Stone Age Crayon Doodle," Nature, September 12, 2018, https://www.nature.com/articles/d41586-018-06664-y.

108. Becky Little, "What Prehistoric Cave Paintings Reveal about Early Human Life," HISTORY, October 5, 2021, https://www.history.com/news/prehistoric-cave-paintings-early-humans.

109. Jérémy Duveau et al., "The Composition of a Neandertal Social Group Revealed by the Hominin Footprints at Le Rozel (Normandy, France)," Proceedings of the National Academy of Sciences of the United States of America 116, no. 39 (September 9, 2019): 19409–14, https://doi.org/10.1073/pnas.1901789116.

110. "First Evidence of Farming in Mideast 23,000 Years Ago," ScienceDaily, July 15, 2015, https://www.sciencedaily.com/releases/2015/07/150722144709.htm.

111. Jonathan Atteberry and Patrick J. Kiger, "10 Amazing Ancient Egyptian Inventions," HowStuffWorks, December 16, 2021, https://science.howstuffworks.com/innovation/inventions/5-amazing-ancient-egyptian-inventions.htm.

112. "The Development of Agriculture," n.d., https://education.nationalgeographic.org/resource/development-agriculture/.

113. "What Is the Sixth Mass Extinction and What Can We Do about It? | Stories | WWF," World Wildlife Fund, n.d., https://www.worldwildlife.org/stories/what-is-the-sixth-mass-extinction-and-what-can-we-do-about-it.

114. "Chinese Coins," Royal Ontario Museum, n.d., https://www.rom.on.ca/en/collections-research/research/world-culture/chinese-coins.

115. Andrew Curry, "Weights Helped Create Europe's First Free Market," Realclearscience.com, June 29, 2021, https://www.realclearscience.com/2021/06/29/weights_helped_create_europes_first_free_market_783379.html.

116. Roxana Bardan, "NASA Confirms DART Mission Impact Changed Asteroid's Motion in Space," NASA, October 11, 2022, https://www.nasa.gov/news-release/nasa-confirms-dart-mission-impact-changed-asteroids-motion-in-space/.

117. Mark Rubin, "Fear of Self-Annihilation and Existential Uncertainty as Predictors of Worldview Defense: Comparing Terror Management and Uncertainty Theories," The Journal of Social Psychology 158, no. 3 (August 31, 2017): 298–308, https://doi.org/10.1080/00224545.2017.1341375.

118. Marissa Moore, "Understanding Terror Management Theory," Psych Central, June 14, 2022, https://psychcentral.com/health/terror-management-theory.

119. Damian Carrington, "Arctic Stronghold of World's Seeds Flooded after Permafrost Melts," The Guardian, May 19, 2017, https://www.theguardian.com/environment/2017/may/19/arctic-stronghold-of-worlds-seeds-flooded-after-permafrost-melts.

120. Tony Ho Tran, "Scientists Are Building a 'Black Box' to Record the End of Civilization," Futurism, December 6, 2021, https://futurism.com/the-byte/earth-black-box-civilization.

121. "Ancient Farmers Spared Us from Glaciers but Profoundly Changed Earth's Climate," ScienceDaily, September 18, 2018, https://www.sciencedaily.com/releases/2018/09/180906141507.htm.

122. Damian Carrington, "Earth's Sixth Mass Extinction Event Under Way, Scientists Warn," The Guardian, July 10, 2017, https://www.theguardian.com/environment/2017/jul/10/earths-sixth-mass-extinction-event-already-underway-scientists-warn.

123. Henry Gee, "Humans Are Doomed to Go Extinct," Scientific American, November 24, 2021, https://www.scientificamerican.com/article/humans-are-doomed-to-go-extinct/.

124. Henry Gee, "Humans Are Doomed to Go Extinct," Scientific American, November 24, 2021, https://www.scientificamerican.com/article/humans-are-doomed-to-go-extinct/.

125. Henry Gee, "Humans Are Doomed to Go Extinct," Scientific American, November 24, 2021, https://www.scientificamerican.com/article/humans-are-doomed-to-go-extinct/.

126. Philip Pullella, "Vatican scientist says belief in God and aliens is OK," Reuters.com, May 14, 2008, https://www.reuters.com/article/us-pope-extraterrestrials-idUKL146364620080514.

127. Giovanni Aquilecchia, "Giordano Bruno | Biography, Death, & Facts," Encyclopedia Britannica, April 9, 2024, https://www.britannica.com/biography/Giordano-Bruno.

128. Brandon Specktor, "Classified UFO Videos Would 'harm National Security' If Released, Navy Says," Livescience.Com, September 10, 2022, https://www.livescience.com/navy-ufo-videos-national-security-threat.

129. Wikipedia contributors, "Emotion," Wikipedia, January 4, 2024, https://en.wikipedia.org/wiki/Emotion.

130. Carter, C Sue, "The biochemistry of love: an oxytocin hypothesis," EMBO reports 14, no. 1 (November 27, 2012): 12–16, https://doi.org/10.1038%2Fembor.2012.191.

131. Jerrold Lee Shapiro, "We Hate What We Fear: Interpersonal Hate from a Clinical Perspective," in Springer eBooks, 2016, 153–77, https://doi.org/10.1007/978-3-319-39277-6_9.

132. "Scientists Identify Brain's 'Hate Circuit,'" ClinicalConnection, October 29, 2008, https://www.clinicalconnection.com/health-news/news-article/2231/scientists-identify-brain-s-hate-circuit.

133. "Brain's 'hate Circuit' Identified," UCL News, October 29, 2008, https://www.ucl.ac.uk/news/2008/oct/brains-hate-circuit-identified.

134. Mariska E. Kret et al., "Emotional Expressions in Human and Non-Human Great Apes," Neuroscience & Biobehavioral Reviews 115 (August 1, 2020): 378–95, https://doi.org/10.1016/j.neubiorev.2020.01.027.

135. Amy Shira Teitel, "Wernher von Braun: History's Most Controversial Figure?," Al Jazeera, May 3, 2013, https://www.aljazeera.com/opinions/2013/5/3/wernher-von-braun-historys-most-controversial-figure/.

136. "Bombing of Hiroshima and Nagasaki - Causes, Impact & Lives Lost," HISTORY, April 18, 2023, https://www.history.com/topics/world-war-ii/bombing-of-hiroshima-and-nagasaki.

137. Zachary Keck, "How Hiroshima and Nagasaki Saved Millions of Lives," The Diplomat, August 07, 2014, https://thediplomat.com/2014/08/how-hiroshima-and-nagasaki-saved-millions-of-lives/.

138. Jamie L. Hanson and Brendon M. Nacewicz, "Amygdala Allostasis and Early Life Adversity: Considering Excitotoxicity and Inescapability in the Sequelae of Stress," Frontiers in Human Neuroscience 15 (June 1, 2021), https://doi.org/10.3389/fnhum.2021.624705.

139. Denise Chow, "Narcissists' Lack of Empathy Detected in Brain Scans," Livescience.Com, June 24, 2013, https://www.livescience.com/37684-narcissistic-personality-disorder-brain-structure.html.

140. Peter Gray, "Why Is Narcissism Increasing Among Young Americans?," Psychologytoday.com, January 16, 2014, https://www.psychologytoday.com/us/blog/freedom-learn/201401/why-is-narcissism-increasing-among-young-americans.

141. Dennis Shen, "A rise in narcissism could be one of the main causes of America's political and economic crises," LSE, June 28, 2017, https://blogs.lse.ac.uk/usappblog/2017/06/28/a-rise-in-narcissism-a-root-of-americas-crisis.

142. Ciara Nugent, "'Abuse Is a Pattern.' Why These Nations Took the Lead in Criminalizing Controlling Behavior in Relationships," TIME, June 21, 2019, https://time.com/5610016/coercive-control-domestic-violence/.

143. Rowan Hooper, "'Spectrum of Empathy' Found in the Brain," New Scientist, September 18, 2006, https://www.newscientist.com/article/dn10108-spectrum-of-empathy-found-in-the-brain/.

ABOUT THE AUTHOR

Mark Christensen grew up in San Francisco Bay area, CA, before moving to Denver, CO, to earn a degree in Architecture. His work experience is diverse enough to make one's head swim ranging from dancing to aviation as a private pilot, snowboarding, and music production. Since 1997, he has been living in Los Angeles and has written, produced, and directed a number of movies, mainly in the science fiction genre.

Mark's endless interest in cosmology, physiology, and sociology combined with his creative work on the sci-fi movie project Smoothies ultimately led him to come up with his Fungus Theory introduced via the book Fungus Theory of Conscious-Growth, which marks his debut as an author. More about Mark and his work is online at https://marklchristensen.com/. Readers can also look at his film portfolio online at https://www.findartfoundation.org/.

ABOUT THE AUTHORS

Made in United States
Orlando, FL
04 November 2026

90200936R00184